STEELE'S NEW CHEMISTRY.

FOURTEEN WEEKS

IN

CHEMISTRY.

BY

J. DORMAN STEELE, PH.D.,

AUTHOR OF THE FOURTEEN-WEEKS SERIES IN NATURAL SCIENCE.

"Bright and glorious is that revelation
Written all over this great world of ours.

LONGFELLOW

A. S. BARNES & COMPANY,

NEW YORK, CHICAGO AND NEW ORLEANS.

THE FOURTEEN WEEKS' COURSES

IN

NATURAL SCIENCE,

BY

J. DORMAN STEELE, A.M., PH.D.

Fourteen Weeks in Natural Philosophy,
Fourteen Weeks in Chemistry,
Fourteen Weeks in Descriptive Astronomy,
Fourteen Weeks in Popular Geology,
Fourteen Weeks in Human Physiology,
Fourteen Weeks in Zoology,
Fourteen Weeks in Botany,
A Key, containing Answers to the Questions and Problems in Steele's 14 Weeks' Courses,

---

A HISTORICAL SERIES,

ON THE PLAN OF STEELE'S 14 WEEKS IN THE SCIENCES.

A Brief History of the United States,
A Brief History of France,

The same publishers also offer the following standard scientific works, being more extended or difficult treatises than those of Prof. Steele, though still of Academic grade.

Peck's Ganot's Natural Philosophy,
Porter's Principles of Chemistry,
Jarvis' Physiology and Laws of Health,
Wood's Botanist and Florist,
Chambers' Elements of Zoology,
McIntyre's Astronomy and the Globes,
Page's Elements of Geology,

PREFACE

TO THE FIRST EDITION.

IN the preparation of this little volume the author lays no claim to originality: his has been the far humbler task of endeavoring to express, in simple, interesting language, a few of the principles and practical applications of Chemistry. There is a large class of pupils in our schools who can pursue this branch only a single term, the time assigned to it in most institutions. They do not intend to become chemists, nor even professional students. If they wander through a large text-book, they become confused by the multiplicity of strange terms, which they cannot tarry to master, and, as the result, too often only "see men as trees walking." Attempts have been made to reach this class by omitting or disguising the nomenclature; but this robs the science of its mathematical beauty and discipline, while it does not fit the student to read other chemical works or to understand their formulæ. The author has tried to meet this want by omitting that which is perfectly obvious to the eye—that which everybody knows already—that which could not be long retained in the memory—and that which is essential only to the chemist. He has not attempted to write a reference-book, lest the untrained mind of the learner should become clogged and wearied with a multitude of detail. He has sought to make a pleasant study which the pupil can master in a single term, so that all its truths may become to him "household words." Botany, Natural Philosophy, and Physiology are omitted, since they are now pursued as separate branches. Unusual importance is given to that practical part of chemical knowledge which concerns our every-day life, in the hope of bringing the school-room, the kitchen, the farm, and the shop in closer relationship. This work is designed for the instruction of youth, and for their sake clearness and simplicity have been preferred to recondite accuracy. If to some young man or woman it becomes the opening door to the grander temple of Nature beyond, the author will be abundantly repaid for all his toil.

NOTICE.—The publication of "Fourteen Weeks in Chemistry," with the *Old Nomenclature*, is STILL CONTINUED. Copies may be obtained of all booksellers, or of the publishers, A. S. BARNES & Co., New York and Chicago.

PREFACE

TO THE REVISED EDITION.

SIX years ago, at the solicitation of his fellow-teachers, the author offered this work to the profession. Having been prepared for the use of his own classes, and embodying his oral instructions, it naturally partook of the peculiarities of that method. The desire was to interest pupils in scientific study. He believed that a chemical fact is no less a truth because made attractive by an imaginative garb. If thus a child could be won to its consideration, the intrinsic beauty of the subject would lure him on, and so at last he would come to pursue it into the labyrinths of dry, technical works.

The hearty reception of the book at once and its constantly increasing sale, the demand for an entire series on the same plan, words of approval from educators whose commendation it was a great satisfaction to have won, the fact that several other series based upon the same general idea have since appeared, and, above all, the assurance that the books have gone into hundreds of schools where science had never been taught before—have convinced the author of the inherent correctness of his view.

A demand having arisen for the admission of the new nomenclature into the book, the opportunity is gladly taken of making such revision as the daily use of the work in the class-room, and the advice of others, have suggested.

The author would here acknowledge his special indebtedness to the many teachers who, sympathizing with his plan of popularizing science, have pointed out what they considered defects in its execution, and given him the benefit of such illustrations and methods as they have found serviceable. The value of these criticisms has been shown in the increased worth of each edition of this series.

The usual authorities have been freely consulted in this revision. The following have been found of especial service: Miller's Elements of Chemistry (4th London Edition), Tomlinson's Miller's Inorganic Chemistry, Roscoe's Lessons in Chemistry (London, 1869), Bloxam's Metals, and Fownes's Manual of Chemistry (London, 1873). In addition, reference has been had to the works of Cooke, Draper, Nichols, Fresenius, Muspratt, Faraday, Watts, Stockhart, Moffit, Gmelin, Griffin, Tyndall, Odling, Noad, Williamson, Wilson, Galloway, Youmans, Regnault, Thomson, Valetin, Gregory, Porter, Will, and many others.

SUGGESTIONS TO TEACHERS.

IT is advised that in the use of this book the topical method of recitation should be adopted. So far as possible, the order of the subjects is uniform—viz., SOURCE, PREPARATION, PROPERTIES, USES, and COMPOUNDS. The subject of each paragraph indicates a question which should draw from the pupil the substance of what follows. At each recitation the scholar should be prepared to explain any point passed over during the term, on the mention of its title by the teacher. Such reviews are of incalculable value. While some are reciting, let others write upon specified topics at the blackboard, after which the class may criticise the thought, the language, the spelling, and the punctuation. *Never allow a pupil to recite a lesson*, or answer a question, except it be a mere definition, *in the language of the book.* The text is designed to interest and instruct the pupil; the recitation should afford him an opportunity of expressing what he has learned, in his own style and words. Every pupil should keep a lecture-book, in which to record under each general head of the text-book all the experiments, descriptions, and general information given by the teacher in class. In order to accustom the scholar to the nomenclature, use the symbols constantly from the beginning: they may seem dull at first, but if every compound be thus named, a familiarity with chemical language will be induced that will be as pleasing as it will be profitable. If time will admit, in addition, have weekly essays prepared by the class, combining information from every attainable source.

Ocular demonstration is absolutely necessary to any progress in the study of chemistry. Simple directions with regard to the experiments are given in the Appendix (see page 245) which will enable the unprofessional chemist to perform them readily, and, in case it is convenient for the pupils to work in the laboratory, will guide them in their investigations. The subject of *Qualitative Analysis* is also explained so clearly, and the directions are so complete (see page 268), that even the amateur student can grasp the subject and demonstrate its principles.

Teachers desiring pleasant information to relieve the recitation hour, will find it in that delightful work of Dr. Nichols, *Fireside Science.* Many curious and entertaining stories and facts are given in a book entitled *Treasures of the Earth.* For a common work of reference, Miller's *Elements of Chemistry*, 3 vols. octavo, will be most generally useful. These books may be obtained of the publishers of this series.

TABLE OF CONTENTS.

I.—INTRODUCTION.

II.—INORGANIC CHEMISTRY.

III.—ORGANIC CHEMISTRY.

IV.—APPENDIX.

I.

Introduction.

"DEAD mineral matter," as we commonly call it, is instinct with force. Each tiny atom is attracted here, repelled there, holds and is held as by bands of iron. No particle is left to itself, but, watched by the Eternal Eye and guided by the Eternal Hand, all obey immutable law. When Christ declared the very hairs of our head to be numbered, he intimated a chemical truth, which we can now know in full to be, that the very atoms of which each hair is composed are numbered by that same watchful Providence.

THE

ELEMENTS OF CHEMISTRY.

INTRODUCTION.

Chemistry treats of the composition of bodies and the specific properties * of matter.—*Examples:* water consists of two gases, hydrogen and oxygen; gold is yellow.

Organic Chemistry deals with those substances which have been produced by life.—*Examples:* flesh and wood. ***Inorganic Chemistry*** is confined to those which have not been formed by life.—*Examples:* sand, glass, metals.

An Element is a kind of matter which has never been separated into other substances.—*Examples:* gold, iron. Sixty-four elements are now known,† fifty-one of which are considered to be metals and thirteen non-metals.

Chemical Affinity is that force which causes the elements of matter to unite and form new compounds. It acts at distances so slight as to be insensible, and upon the most dissimilar substances: the more dissimilar, the

* See the Introduction to Physics for definitions of these terms.

† It is not probable that the list is complete, but we cannot suppose that any very abundant element is yet to be found. Indeed, of those made known since 1774 (the year of the discovery of O, Cl, etc.), the majority are only chemical curiosities.—The division into metals and non-metals is an arbitrary one, and not clearly defined. (See note, p. 123.)

stronger the union.—*Example:* a little potassium chlorate * and sulphur mixed in a mortar will not combine, but a slight pressure of the pestle will bring them within the range of attraction, when they will burn with a loud explosion. Nothing in the nature or appearance of an element indicates its chemical affinity, and it is only by trial that we can tell with what it will combine. This attraction is not a mere freak of nature, but a force imparted to matter by God himself for wise and beneficent purposes.

Compounds, in their properties, are in general very unlike their elements.†—*Examples:* yellow sulphur and white quicksilver form red vermilion; inert charcoal, hydrogen, and nitrogen produce the deadly prussic acid; solid charcoal and sulphur make a colorless liquid; poisonous and offensive chlorine combines with the brilliant metal sodium to form common salt.

Heat and ***Light*** favor chemical action, and frequently develop an affinity where it seemed to be wanting. The former especially, by its expansive force, tends to drive the elements of a compound without the range of old attractions and within that of new ones.—*Examples:* gun-cotton, when lying in the air, is apparently harmless, but a spark of fire will produce a brilliant flash, and cause it to disappear as a gas: nitrate of silver in contact with organic matter turns black, by the action of the light.

Solution also aids in chemical change, as it destroys cohesion and leaves the atoms free to unite.—*Example:* sodium carbonate ‡ and tartaric acid mixed in a glass

* "Chlorate of potash."

† "The elements have no more likeness to the compounds which they form than the separate letters of the alphabet have to the words which may be made from them."—MILLER.

‡ "Carbonate of soda."

will not combine, but the addition of water will cause a violent effervescence.

Nomenclature.—The elements which were known anciently retain their former names. Those discovered more recently are named from some peculiarity.—*Examples:* chlorine, from its green color; bromine, from its bad odor. The uniform termination *um* has been given to the lately found metals.—*Examples:* potassium, sodium. A similarity of ending in non-metallic elements indicates some analogy.—*Examples:* silicon, boron; iodine, bromine.

Symbols.—For the sake of brevity chemists use a kind of short-hand. The first letter of its English name is generally taken as the symbol of an element. When that would produce confusion, the Latin initial is substituted, and in some cases a second letter added.—*Examples:* carbon and chlorine both commence with C; so the latter takes Cl for its symbol. Silver and silicon both begin with Si, hence the former assumes Ag, from its Latin name, Argentum. If more than one atom of an element be used in forming a molecule of a compound, this is shown by writing the number below the symbol.—*Example:* H_2O indicates that in a molecule of water there are two atoms of hydrogen and one of oxygen.

The Atomic Weight of an element expresses the proportion by weight in which it unites with other elements. There is no chance-work in nature. No matter under what circumstances a compound is formed, the proportion of its elements is the same.—*Example:* the carbonic acid produced amid the roar of a conflagration or the explosion of a volcano is identical with that made in the quiet burning of a match.

In the table on page 288 are given the symbols of the elements, their atomic weights, etc. The metals are printed in Roman letters and the non-metals in *italics*. Hydrogen is taken as the unit of atomic weight.

Constitution of Bodies.*—All bodies are considered to be collections of molecules, as molecules † are of atoms. In some of the elementary substances the molecule and the atom are identical, but in many kinds of matter several atoms are combined to form a molecule. Thus the molecules of nitrogen, chlorine, and oxygen, contain two atoms each, and arsenic and barium four.‡ The molecule of a compound always contains two or more atoms. The molecule is destructible, but the atom is indestructible. The specific properties of a substance reside in the molecule, and to break up the molecule is to destroy the substance. A physical change goes no further than the molecule, and hence does not affect the integrity of a substance; but a chemical change consists in a re-arrangement of the atoms and the formation of new molecules, and hence, of new properties. Whenever a combination then takes place it is by elementary atoms and not by molecules. In Physics we learned how physical changes are caused by transformations of force without loss of energy. In Chemistry we shall see how chemical changes are produced by transformations of atoms without loss of matter.

* The subject of the quantivalence of atoms is treated on page 142, as some knowledge of the chemical behavior of atoms is necessary fully to understand its force.

† A molecule is the smallest particle of a substance which can exist in a separate form. It is the unit of the philosopher as the atom is of the chemist.

‡ See the table in Appendix, page 288, under Molecular Weight.

The ***Atomic Theory*** which lies at the basis of chemistry, as now understood, supposes:

1. That each element is composed of indivisible* atoms which are exactly equal in size and weight.

2. That the atomic weights represent the relative weights of the atoms of various kinds.

3. That compounds are formed by the union of atoms of different kinds in the proportion of their atomic weights or multiples of them.

4. That the molecular weight of a compound is equal to the sum of the atomic weights of its elements.

A ***Binary Compound*** is a union of two elements. In writing its symbol, we place first that of the electro-positive,† and then that of the electro-negative element. In writing the name, or in reading the symbol, the latter element takes the termination *ide.* Thus potassium and iodine form the compound which is written KI, and read potassium iodide ; sodium and chlorine, NaCl, sodium chloride; zinc and oxygen, ZnO, zinc oxide.‡

One atom of O in a molecule forms the monoxide or protoxide, two the dioxide or binoxide, three of O and two of the other element, the sesquioxide (meaning $1\frac{1}{2}$), and the highest number, the peroxide. Thus,

* The atoms are termed indivisible, as the chemist cannot break up an atom of O any more than the astronomer can the planet Jupiter. Yet there is no more reason for supposing the elementary atom incapable of division by processes now unknown than that Jupiter cannot be resolved into smaller masses.

† See these terms defined under Electricity in Fourteen Weeks in Physics.

‡ There are three variations from this statement which should be noticed. 1. Many chemists give the electro-positive element the termination *ic*, thus reading KI, potassic iodide. 2. The electro-negative element is often read first, and the word *of* placed between the elements ; thus KI is called the iodide of potassium. 3. In the case of phosphorus, carbon, and sulphur, the termination *uret* is sometimes used instead of *ide* ; thus FeS is read the sulphuret of iron, instead of iron (ferrous) sulphide or the sulphide of iron.

N_2O	=	Nitrogen	Monoxide (protoxide),
N_2O_2	=	"	Dioxide (binoxide),
N_2O_3	=	"	Trioxide,
N_2O_4	=	"	Tetroxide,
N_2O_5	=	"	Pentoxide.

Acids, Bases, and Salts.—There are two large classes of oxides chemically opposed to each other, termed *acids* and *bases;* their compounds are called *salts.*

The ***Acids*** are generally sour * and turn vegetable colors—such as the infusion of blue litmus, or of purple cabbage †—to a bright red. They are named from the elements with which O combines. The termination indicates the amount of O,—*ic* representing the greater, and *ous* the lesser.—*Example:* sulphur forms two acids of different strength—sulphuric, the stronger, and sulphurous, the weaker. If an acid has been afterwards found containing more O than the stronger, it takes the prefix *per;* one having less than the weaker, the prefix *hypo.*—*Example:* chlorine combines with oxygen and hydrogen to form a series of acids in regular gradation.

Hypochlorous	Acid	$HClO$,
Chlorous	"	$HClO_2$,
Chloric	"	$HClO_3$,
Perchloric	"	$HClO_4$.

Hydrogen by its union with different elements forms acids which contain no O. These combine the names of

* Certain acids, as well as certain bases, are insoluble in water, and hence have no taste. They, however, combine to form salts, which is their true test.

† Paper tinged blue with a solution of litmus (a coloring matter obtained from certain lichens) should be constantly at hand in the laboratory, to determine the presence of a free acid. The same paper faintly reddened by vinegar, or any other acid, is a convenient test for the alkalies. The cabbage solution is made by steeping red cabbage-leaves in water, and straining the purplish liquid thus obtained.

both elements.—*Example:* hydrogen and chlorine form hydrochloric acid.

The ***Bases*** are commonly oxides of the metals. Their termination, as in the acids, indicates the amount of oxygen. Thus mercury has two oxides, HgO and Hg_2O, termed respectively mercuric oxide and mercurous oxide; iron forms FeO, ferrous, and Fe_2O_3, ferric oxide. *The alkalies** are bases which are soluble in water, have a soapy taste and feel, turn red litmus to blue, and red-cabbage solution to green, neutralize the acids and restore the colors changed by them. *The property which the acids and bases thus have of uniting with each other and destroying the chemical activity which either possesses alone, is their distinguishing trait.*†

The ***Salts*** are compounds formed by the union of an acid and a base.‡ In naming a salt, the termination of the acid is changed—an *ic* acid forming an *ate* compound, and an *ous* acid an *ite* compound. Thus the salts of sulphuric acid are called sulphates, and of sulphurous acid, sulphites; of nitric acid, nitrates, and of nitrous acid, nitrites. Sulphuric acid combining with ferrous oxide forms ferrous sulphate, and with ferric oxide, ferric sulphate.

A ***Formula*** is an algebraic statement of the symbols and relations of several compounds. The sign + in-

* The alkalies are compounds of H, O and a metal. They are hence called *hydroxides:* as KHO (potassium hydrate, caustic potash), NaHO (sodium hydrate).

† To a part of the purple-cabbage solution add a few drops of a solution of caustic potash: a green liquid will be produced. To another portion add a few drops of sulphuric acid: the solution will become red. Pour the red acid liquor into the green alkaline one, and stir the mixture: the red color at first disappears, and the whole remains green; but on adding it cautiously, a point is reached at which it assumes a clear blue color. There is then no excess of acid or alkali; and on evaporation, a neutral salt, *potassium sulphate*, may be obtained.

‡ We shall come hereafter to substitute for this simple definition the more exact chemical one, that a salt is an acid in which one or more atoms of H have been replaced by a metal. (See note, p. 44: reaction, p. 51: and notes, p. 128.)

dicates a feeble attraction or a mere mixture. The sign = indicates conversion into. The comma or the period denotes a combination. The brackets and coefficients are used as in algebra.

Mathematics of Chemistry.—There is a Divine law of harmony which runs like a golden thread through all nature, giving always unity and completeness. Its beauty and simplicity are nowhere seen more clearly than in the law of atomic weights. Applying the fourth principle of the atomic theory, we see that the atomic weight of any element in a compound, divided by the molecular weight of that compound, is the proportion of that element contained in it.—*Example:* the molecular weight of water, H_2O, is $2+16=18$; hence the proportion of H is $\frac{2}{18}$ or $\frac{1}{9}$, and of O, $\frac{16}{18}$ or $\frac{8}{9}$. In 10 lbs. of H_2O, there are therefore $10\times\frac{8}{9}$ or $8\frac{8}{9}$ lbs. of O, and $10\times\frac{1}{9}$ or $1\frac{1}{9}$ lbs. of H.*

Apply this principle to the solution of the following

PROBLEMS.

1. In a 25-lb. sack of table salt (NaCl, sodium chloride), how many lbs. of the metal sodium ?†
2. In 14 lbs. of iron-rust (Fe_2O_3), how much O ?
3. How much S is there in 2 lbs. of SO_2 ?
4. How much S is there in 2 lbs. of H_2SO_4 (sulphuric acid)?
5. How much O is there in 5 lbs. of HNO_3 (nitric acid)?
6. How much H is there in 6 lbs. of HCl (hydrochloric acid)?
7. How much potash (K_2O) could be made from 3 lbs. of potassium (K)? ‡

* This may also be solved by the following proportion :—
The atomic weight of an element : molecular weight of a compound :: weight of the element : weight of the compound.

$2 : 18 :: x : 10$ lbs. $x=1\frac{1}{9}$ lbs. (H).
$16 : 18 :: x : 10$ lbs. $x=8\frac{8}{9}$ lbs. (O).

† $23 : 58.5 :: x : 25$ lbs. $x=9\frac{97}{117}$ lbs. (Na).

‡ $78 : 94 :: 3$ lbs. $: x$. $x=3\frac{8}{13}$ lbs. (K_2O).

II.

Inorganic Chemistry.

"IN the de-oxidation and re-oxidation of the hydrogen in a single drop of water, we have before us, so far as force is concerned, an epitome of the whole of life."—HINTON.

INORGANIC CHEMISTRY.

THE NON-METALS.

OXYGEN.

Symbol, O....Atomic Weight, 16....Specific Gravity, 1.1.

THE name Oxygen means acid-former, and was given because it was supposed to be the essential principle of all acids; but hydrogen has since been found to be the true acid-maker.

Fig. 2.

Collecting O over a pneumatic tub.

Source.—O is the most abundant of all the elements—comprising by weight $\frac{1}{5}$ of the air, $\frac{8}{9}$ of the water, $\frac{3}{4}$ of all animal bodies, and about $\frac{1}{2}$ of the crust of the earth.

Preparation.—The simplest method of preparing O for experimental purposes is to heat a mixture of potassium chlorate ($KClO_3$) and manganese dioxide (MnO_2) * in a flask, and collect the gas over a pneumatic tub, as in the accompanying illustration.†

The ***Reaction,*** chemical change, is as follows:

$$2\,KClO_3 \longrightarrow 2\,KCl + O_6$$

Two molecules of potassium chlorate are converted into two of potassium chloride and six atoms of O, which pass off as a gas. The reaction may also be represented thus:

(Potassium Chlorate)		(Potassium Chloride)		(Oxygen)
$2\,K\,Cl\,O_3$	=	$2K\,Cl$	+	O_6
$2(39+35.5+3\times16)$		$2(39+35.5)$		6×16
245		149		96
		245		

The O set free will be equal to $\frac{96}{245}$ of the potassium chlorate used, i. e., every 245 parts by weight (grs., oz., or lbs.) will yield 96 parts (grs., oz., or lbs.) of O, and 149 parts (grs., oz., or lbs) of KCl.

A ***Curious Fact*** appears in this process. If the $KClO_3$ were heated alone, a very high temperature would

* This substance is commonly known as binoxide of manganese, and, because of its color, the black oxide of manganese.

† See Appendix for directions in performing this and other experiments.

be necessary, which would liberate the gas rapidly, and often with explosive violence. If, however, we mix with it a little manganese dioxide, the gas will be set free at a much lower temperature, and may be regulated so as to come off a bubble at a time. At the conclusion of the process, the MnO_2 will be found unchanged. The reason of this wonderful action is beyond our comprehension. The influence of one body over another, by its mere presence, is called *catalysis.*

Properties.—O has no odor, color, or taste. It combines with every element except fluorine. From some of its compounds it can be set free by the stroke of a hammer, while from others it can be liberated only by the most powerful means. Its union with a substance is called *oxidation,* and the product an *oxide.* It is a vigorous supporter of combustion.

The following experiments will illustrate its chemical energy.

Fig. 3.

A candle in O.

1. By blowing quickly upward upon a candle, extinguish the flame, and leave a glowing wick. If this be plunged into a jar of O, the coal will burst into a brilliant blaze. The experiment may be repeated many times before the O will be exhausted. A new colorless gas, CO_2, called carbonic anhydride * ("carbonic acid") is formed by the combustion.

* An *anhydride* (without hydrogen) is a substance which, when dissolved in H_2O will unite with its elements and form an acid. It is then strictly a salt in which H plays the part of a base, and is called a hydride. In this state only is it

Fig. 4.

A watch-spring in O.

2. Straighten a watch-spring by drawing it between the fingers. Pass one end through a cork, heat the other slightly and dip it into powdered sulphur. Light this and plunge it into a jar of O, holding it in the neck by the cork. The burning sulphur will ignite the steel, which will burn with a shower of fiery stars, while melted globules of the black oxide of iron (Fe_3O_4) will fall upon the plate below.

Fig. 5.

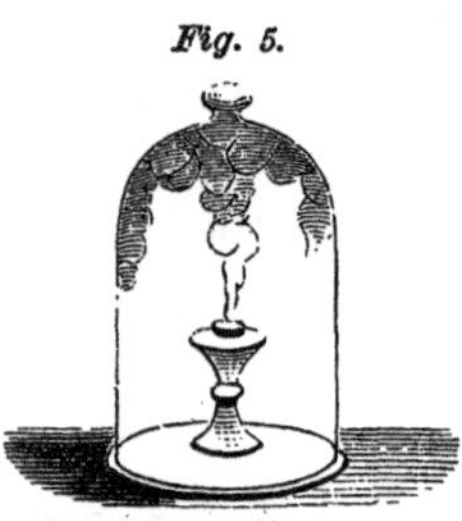

Sulphur in O.

3. Ignite a bit of sulphur placed on a stand, and invert over it a jar of O: it will burn with a beautiful blue light, and the fumes of sulphurous anhydride, SO_2 ("sulphurous acid") will circle about the receiver in curious concentric rings. The gas has a pungent odor, and will be absorbed by the water on the plate, where it may be tested.

4. Place in the bottom of a "deflagrating spoon," ee Appendix, p. 248) a little fine, dry chalk; then wipe a bit of phosphorus, about the size of a pea, very carefully and quickly between pieces of blotting-paper; lay this upon the chalk, and, holding the spoon over a large

properly termed an acid. The two compounds are frequently distinguished as anhydrous (without water), and hydrated (with water). Until of late it has been customary to apply the term acid to either form. Thus CO_2 (carbon dioxide) is strictly carbonic anhydride, but has been so long known as carbonic acid that its proper appellation is rarely used. The same is true of the acids named in the 3d. and 4th. experiments.

Fig. 6.

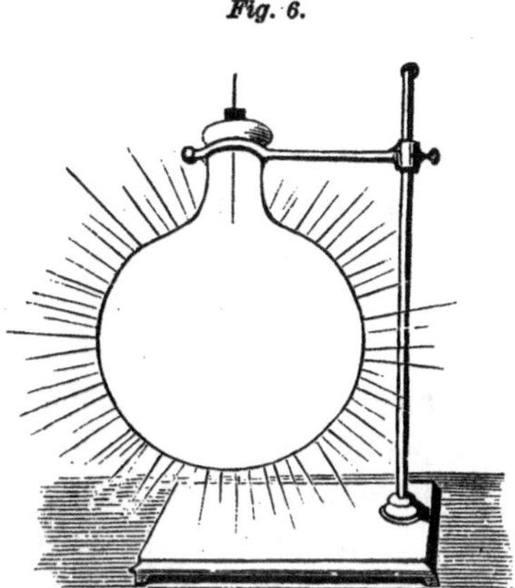

Phosphorus in O. "*The phosphoric sun.*"

jar of O, ignite the phosphorus with a heated wire, and lower it steadily into the gas. The phosphorus will burst into a flood of blinding light, while dense fumes of phosphoric anhydride, P_2O_5, ("phosphoric acid") will roll down the sides of the jar.

5. Make a little tassel of zinc-foil, tip the ends with sulphur as in the 2d experiment, ignite and lower into a jar of O. It will burn with a dazzling light, forming zinc oxide (ZnO).

6. If a piece of charcoal-bark be ignited and lowered into a jar of O, it will deflagrate with bright scintillations.

The Destructive Agent of the Air.—O is the active principle of the atmosphere. Comprising one-fifth of the common air, it is ever-present, and ever-waiting. We gather a basket of peaches and set them aside. In a short time, black spots appear, and we say they are decaying. It is only the O corroding them, *i. e.*, breaking up their chemical structure to form new and unpleasant compounds. To prevent this action, we place the fruit in a can, heat it to expel the O, and seal it tightly.—We open the damper of the stove and the air rushes in. The O immediately attacks the heated fuel. Every two atoms combine with an atom of C and fly off into the air as CO_2.—We cut a finger, and soon feel the O at work upon the quivering nerve beneath. We apply a strip of "court-plaster" to keep out the air and give nature an opportunity to heal the wound.*—Our teeth decay only because of the action of the O. The dentist saves them by filling any break in the enamel with a cement which is already oxidized, or with a metal, as Au or Pt, which has little affinity for O.—The H_2O in the cistern becomes foul and putrid. We uncover it; in rushes the O, picks up each atom of impurity, and sinks to the bottom. The thick sediment we find when it is cleaned in the spring, is but the ashes of this combustion.†—The blacksmith draws a red-hot iron from his forge. While the metal is glowing, the O forms scales of the black oxide of iron (Fe_3O_4), which fly blazing in every direction.‡—We wipe our knives

* The treatment of a burn as well as a cut consists in the immediate exclusion of the air. It is a mistake to suppose that a salve will "draw out the fire" of a burn, or heal a bruise or cut. The vital force must unite the divided tissue by the deposit of material, and the formation of new cells. (*Physiology*, p. 205.)

† "As the vessel sets sail from London, the captain fills the water-casks with water from the River Thames, foul with the sewage of the city, and containing 23 different species of animalcules; yet, in a few days, the O contained in the air dissolved by the H_2O, will have cleansed it, and the H_2O will be found sweet and wholesome during the voyage."

‡ Quite in contrast to this pyrotechnic display is the action of the O upon the Fe contained in writing-fluid. At first the words are pale and indistinct, but in a few hours the O, noiselessly combining with the metal (see p. 212), brings out every letter in clear, bold characters upon the page.

and forks, and lay them carefully away; but if we have left on them a particle of moisture, since H_2O favors chemical change, the O will find it, and corrode the steel.*—An animal dies, and the O at once begins to remove the body. The atoms which have been used to perform the functions of life, are separated by the O, and set at liberty to enter into new combinations.

O *in the Human System.*—We take the air into our lungs. Here the blood† absorbs the O, and bears it to all parts of the body, depositing it wherever it is needed. Laden with this life-giving element, the vital fluid sweeps tingling through every artery and vein, distends each capillary tube, sends the quick flush to the cheek, combines with a portion of the food thrown into the circulation from the stomach, breaks up every worn-out tissue, burns up the muscles, and sets free their force, until at last it comes back through the veins dark and thick with the products of the combustion—the cinders of the fire within us.

Combustion and Heat.—All ordinary processes of fermentation, decay, putrefaction, and fire, are produced by a union of O with a substance, and are only different forms of oxidation. They differ in the time employed in the operation. If O unites rapidly,

* The compound here formed will be a higher oxide than that produced at the blacksmith's forge, since a portion of the O which there united with the iron was driven off by the heat. It will be the red oxide of iron (Fe_2O_3, ferric oxide), or common iron rust, as we see it on stoves and other utensils.

† The blood is full of red corpuscles or cells containing Fe. These are so tiny, that a million of them cluster in the drop which will cling to the point of a needle. Quickly assuming a tawny hue, like the decayed leaves of autumn, they change so rapidly that 20,000,000 perish with every breath.—DRAPER.

These cells when fresh act like little gas-bags in carrying the O through the body.

we call it fire; if slowly, decay. Yet the process and the products are the same. A stick of wood is burned in the stove, and another rots in the forest, but the chemical change is identical. In the combination of an atom of O, a certain amount of heat is produced.* Hence, the house that decays in fifty years, gives out as much heat during that time as if it had been swept off by a fierce conflagration in as many minutes.

The Human Furnace.—The body is like a stove in which fuel is burned, and the chemical action is precisely like that in any other stove. This combustion produces heat, and our bodies are kept warm by the constant fire within us. We thus see why we fortify ourselves against a cold day by a full meal. When there is plenty of fuel in our human furnaces, the O burns that; but if there is a deficiency, the destructive O must still unite with something, and so it combines with the flesh;—first the

* "When considerable masses of iron are allowed to rust, a distinct elevation of temperature is often perceived. This is seen when a heap of iron turnings of from 10 lbs. to 20 lbs. is moistened with water and exposed to the air. A curious illustration of the fact was afforded during the manufacture of the Mediterranean Electric Cable. The copper conducting wire of this cable was coated with gutta-percha; this was covered with a serving of tar and hemp, and the whole was enclosed in a strong casing of iron wire. The cable as it was manufactured was coiled in tanks filled with water. These tanks leaked, and the water was therefore drawn off, leaving a quantity of cable, about 163 nautical miles in length, coiled into a mass about 30 feet in diameter with an eye or central space of 6 feet; the height of the coil was about 8 feet. Rapid oxidation took place, and the temperature at the centre of the coil, nearly three feet from the bottom, rose in four days from 66° to 79°, although the temperature of the air did not exceed 66° during the period, and was as low as 59° part of the time. In other parts of the mass the heat rose so high as to cause the water to evaporate sufficiently rapidly to produce a visible cloud of vapor, and to give rise to apprehensions that the insulating power of the cable would be destroyed by the softening of the gutta-percha. No doubt the rise of temperature would have been still greater had it not been checked by the affusion of cold water; but the oxidation and the heating were renewed when the cooling was discontinued. The oxidation occurred only on the external surface of the iron wires, that portion in contact with the tarred hemp remaining perfectly bright."—MILLER.

fat, and the man grows poor; then the muscles, and he grows weak; finally the brain, and he becomes crazed. He has simply burned up, as a candle burns out to darkness.

O *Produces Motion.*—As soon as we begin to perform any unusual exercise, we commence breathing more rapidly,—showing that, in order to do the work, we need more O to unite with the food * and muscles. In very violent labor, as in running, we are compelled to open our mouths, and take deep inspirations of O. This increased fire within elevates the temperature of the body, and we say "we are so warm that we pant." Really it is the reverse. The panting is the cause of our warmth.

During sleep the organs of the body are mostly at rest, except the heart. To produce this small muscular exertion very little O is required. As our respiration is, therefore, slight, our pulse sinks, the heat of our body falls, and we need much additional clothing to keep warm.† Thus we require O not only to keep us warm, but also to do all our work. Cut off its supply, and we grow cold; the heart struggles spasmodically for an instant, but the motive power is gone, and we soon die.

How O *gives us Strength.*—Our muscles, as well as the food from which they are formed, consist of com-

* It is probable that a portion of our food, especially the carbonaceous, is oxidized directly without becoming an integral part of the body. The heat thus set free by the principle of the *Conservation of Energy* (see *Physiology*, page 134), may be converted into muscular force.

† Animals that hibernate show the same truth. The marmot, for instance, in summer is warm-blooded; in the winter its pulse sinks from 140 to 4, and its heat corresponds. The bear goes to his cave in the fall, fat; in the spring he comes out lean and lank. Cold-blooded animals have very inferior breathing apparatus. A frog, for example, has to swallow air by mouthfuls, as we do water. Others have no lungs at all, and breathe in a little air through the skin, enough barely to exist. Is it strange they are cold-blooded?

plex organic bodies, and the tension of the pent-up force is very great. Thus in flesh, starch, sugar, etc., the molecules are very large (see p. 100), and, when these oxidize into the smaller ones of water, carbonic acid, and ammonia, the hidden energy thus liberated gives us heat and strength.* It is merely the transference of force from one organic body to another. One decays, the other grows. One drops in the scale of life, the other rises. One loses as the other gains. As no matter is either lost or gained in any chemical change, so also no force is lost or gained, but all must be accounted for. Action and reaction are equal in chemistry as in philosophy.

The Burning of the Body by O.—A man weighing 150 lbs. has 64 lbs. of muscle. This will be burned in about 80 days of ordinary labor. As the heart works day and night, it burns out in about a month. So that we have a literal "new heart" every thirty days. We thus dissolve, melt away in time, and only the shadow of our bodies can be called our own. They are like the flame of a lamp, which appears for a long time the same, since it is "ceaselessly fed as it ceaselessly melts away." The rapidity of this change in our bodies is remarkable. Says Dr. Draper: "Let a man abstain from water and food for an hour, and the balance will prove he has become lighter." This action of O, so destructive—wasting us away constantly from birth to death, is yet essential to our existence. Why is this? Here is the glorious paradox of life. *We live only as we die.* The moment we cease dying, we cease living. All our life is produced

* This latent force is called a potential one, and the same force, when sensible, is termed a dynamic one. In the former case it is hidden and ready to burst out at any time; in the latter, it is in full action. Potential force is contained in the powder of a loaded gun. Dynamic force drives the bullet to the mark.

by the destruction of our bodies. No act can be performed except by the wearing away of a muscle. No thought can be evolved except at the expense of the brain. Hence the necessity for food to supply the constant waste of the system,* and for sleep to give nature time to repair the losses of the day. Thus, also, we see why we feel exhausted at night and refreshed in the morning.

O *the Common Scavenger.*—God has no idlers in his world. Each atom has its use. There is not an extra particle in the universe. The mission of oxygen, so destructive in its action, is therefore essential, that every waste substance may be collected and returned to the common stock, for use in nature's laboratory. In performing this general task, its uses are most important and necessary. It sweetens water, it keeps the avenues of the body open and unclogged,† it preserves the air wholesome. It becomes, in a word, the universal scavenger of nature. Every dark cellar of the city, every recess of the body, every nook and cranny of creation, finds it waiting; and the instant an atom is exposed, the oxygen seizes upon it. A leaf falls, and the O forthwith commences its destruction. A tiny twig, far out at the end of a limb, dies, and the O immediately begins its removal. A pile of decaying vegetables, a heap of rubbish, the dead body of an animal, a fallen tree, the houses we build for

* This food must be organic matter endowed with potential power treasured up in the plant. When it is transformed into flesh, perhaps made still more vital in the process, we have this force standing ready to be used again at our pleasure. When we will it, the O combines with the flesh and sets free the energy for us to apply.

† Huxley very prettily calls O, in this connection, the "great sweeper" of the body, since it lays hold of all the waste matter of the system, and burning it up, removes it out of the way.

our shelter, even the monuments erected above our final resting-place, are all gnawed upon by what we call the "insatiate tooth of time." It is only the constant corrosion of this destructive agent—oxygen.

Consumption of O.—Each adult uses daily $1\frac{1}{2}$ lbs. of O. The combustion of 1 lb. of coal requires $2\frac{2}{3}$ lbs. of O: so that the ship which burns 1,000 tons in crossing the ocean, takes out of the air 2,666 tons of O. Supposing the population of the earth to be 1,200,000,000, and each person to consume 1 lb. of O; adding as much more to sustain fires; twice as much for the wants of animals, and four times as much for the varied processes of decay, the daily consumption of O reaches the enormous sum of 4,800,000 tons (Faraday). Yet the atmosphere contains over one quadrillion tons, and even this vast aggregate is a mere fraction compared with the O locked up in the ocean and the rock.

Results if the Air were Undiluted O.—The fire element would run riot everywhere. Metal lamps would burn with the oil they contain. Our stoves would blaze with a shower of sparks. A fire once kindled would spread with ungovernable velocity, and a universal conflagration would quickly wrap the world in flame.

OZONE.

Ozone is an *allotropic* form of O—*i. e.*, a form in which the element itself is so changed as to have new properties.

Source.—It is always perceived during the working of an electric machine, and is then called "the electric smell." It has also been detected near objects just struck by lightning. Electricity is supposed to have something to do with the formation of the ozone in the atmosphere.

Preparation.—Pour a little ether into a jar of common air, and stir in its vapor a heated glass rod. The O will be immediately changed into its allotropic form—ozone—which can be recognized by its pungent odor. It may also be tested by a paper wet with a mixture of starch and potassium iodide (KI). The ozone sets free the iodine, which unites with the starch, forming blue iodide of starch.* At a temperature a little above that of boiling water, the ozone will turn into O.

Fig. 7.

Preparing ozone.

Properties.—Ozone is still more corrosive than oxygen. It bleaches powerfully, and is a rapid disinfectant. A piece of tainted meat plunged into a jar of it is instantly deodorized, and it is probable that, even in minute quantities, this gas exercises a powerful influence in purifying the atmosphere. Its over-abundance in the air is supposed to produce influenzas, diseases of the lungs, etc., and its absence to cause fevers, agues, and kindred diseases.

* If a piece of the dry iodized paper be exposed upon a clear day to the open air of the country, in a few minutes it will assume a bluish tint. In cloudy, foggy weather, or in cities, this effect is rarely observed.—Recent experiments have thrown doubt on the former tests for the abundance of ozone in the atmosphere, as well as the existence of antozone. Further researches are demanded.

Antozone (the opposite of ozone) is always formed at the same time as ozone, but returns to ordinary O more readily. Its distinguishing trait is its tendency to form clouds with O. We notice it in the oxidation of phosphorus, as a white mist which remains long after the phosphorus oxides have been dissolved by the H_2O. The gray smoke that lingers around chimneys, steam-engines, etc., is composed of antozone.

PRACTICAL QUESTIONS.

1. Are all acids sour?
2. What is the difference between an *ate*, an *ite*, and an *ide* compound?
3. Why does not canned fruit decay?
4. Where is the higher oxide formed, at the forge or in the pantry?
5. Why is the blood red in the arteries, and dark in the veins?
6. Do we need more O in winter than in summer?
7. Which would starve sooner, a fat man or a lean one?
8. How do teamsters warm themselves by slapping their hands together?
9. Could a person commit suicide by holding his breath?
10. Why do we die when our breath is stopped?
11. Why do we breathe so slowly when we sleep?
12. How does a cold-blooded animal differ from a warm-blooded one?
13. Why does not the body burn out like a candle?
14. Do all parts of the body change alike?
15. What objects would escape combustion if the air were undiluted O?
16. How much O can be obtained from 6 oz. of $KClO_3$?
17. How much $KClO_3$ would be needed to produce 2 lbs. of O?
18. How much KCl would be formed in preparing 1 lb. of O?
19. Name a substance from which the O can be set free by a stroke of the hammer.
20. Name one from which the O can only be liberated with extreme difficulty.

21. Is it probable that all the elements are discovered?
22. Is heat *produced* by oxidation?
23. What is the difference between dynamic and potential force?
24. Why does running cause panting?
25. How does O give us force?
26. Does the plant *produce* force?
27. If we burn an organic body in a stove it gives off heat; in the body it produces also motion. Explain.
28. In preparing N, a thin white cloud remains in the jar for a long time. What is it?

NITROGEN.

Symbol, N....Atomic Weight, 14....Specific Gravity, 0.97.

THIS gas is called nitrogen because it exists in nitre.

Sources.—N forms $\frac{4}{5}$ of the atmosphere, and is found abundantly in ammonia, nitric acid, flesh,* and in such vegetables as the mushroom, cabbage, horse-radish, etc. It is an essential constituent of the valuable medicines, quinine and morphine, and of the potent poisons, prussic acid and strychnine.

Fig. 8.

Preparing N.

Preparation.—As the air consists of N and O, the easiest method of obtaining the former gas is to remove the latter. Place in the centre of a deep dish of water a little stand several inches in height, on which a bit of phosphorus may be laid and ignited. As the fumes of phosphoric anhydride ascend, invert a receiver over the stand. The phosphorus will

* Its compounds give to burnt hair and woolen their peculiar odor.

consume the O of the air contained in the jar, leaving the N. Add more H_2O as that in the plate rises. It should occupy $\frac{1}{5}$ of the receiver. The jar will at first be filled with white fumes (P_2O_5), but they will be absorbed by the H_2O in a short time.

Properties.—All descriptions of N are of a negative character. It neither burns nor permits anything else to burn. It neither supports life nor destroys it. Yet a candle will not burn in it, and a person cannot breathe it alone and live, simply because it shuts off the life-giving O. So will a person drown in H_2O, not that the water poisons him, but because it fills his mouth, and shuts out the air. N has only a weak affinity for any of the elements. The instability of its compounds is a striking peculiarity. It will unite with iodine, for example, but a brush with a feather, or a heavy step on the floor will set it free.*

Uses.—*Relation of* N *to Organic Substances.* —Four-fifths of each breath that enters our lungs is N; yet it comes out as it went in,† while that portion of the O which remains behind performs its wonderful work within our bodies. One-fifth of our flesh is N, yet none of it comes from the air we breathe. We obtain all our supply from the lean meat and vegetables we eat. Plants breathe the air through the leaves—their lungs; yet they

* "Like a half-reclaimed gypsy from the wilds, it is ever seeking to be free again; and not content with its own freedom, is ever tempting others, not of gypsy blood, to escape from thraldom. Like a bird of strong beak and broad wing, whose proper place is the sky, it opens the door of its aviary, and rouses and flutters the other and more peaceful birds, till they fly with it, although they soon part company."—*Edinburgh Review.*

† There is a constant exhalation of N through the pores of the skin. This small amount is perhaps absorbed in the lungs, but it is of no use to the body, so far as known.

do not appropriate any of the N obtained in this way, but rely upon the ammonia and the nitric acid their roots absorb from the soil. N enters the stove with the O —the latter unites with the fuel; but the former, having no chemical attraction, passes out of the chimney. Even from a blast furnace, where Fe melts instantly like wax, N comes forth without the smell of fire upon it (p. 150). So inert is it, that it will not unite directly with any organic substance. We must all, animals and plants, depend upon finding it already combined in some chemical compound, and so appropriate it to our use. But even then we hold it very loosely indeed. The tendency of flesh to decompose is largely owing to the instability of the N in its composition.

Difference between* N *and* O.*—We see now how different N is from O. The one is the conservative element, the other the radical. But notice the nice planning shown in the adaptation of the two to our wants. O, alone, is too active, and must be restrained; N, alone, is sluggish, and only fit to weaken a stronger element. Were the air undiluted O, our life would be excited to a pitch of which we can scarcely dream, and would sweep through its feverish, burning course in a few days; were it undiluted N we could not exist a moment. Thus we see that, separately, either element of the air would kill us, O by excess and N by lack of action.

O *and* N *combined*.—A mixture of fiery O and the inert N gives us the golden mean. The O now quietly

* The difference between these two gases can be best illustrated by having a jar of each, and rapidly passing a lighted candle from one to the other; the N will extinguish the flame, and the O relight the coal. By dextrous management, this may be repeated a score of times.

burns the fuel in our stoves and keeps us warm; combines with the oil in our lamps and gives us light; corrodes our bodies and gives us strength; cleanses the air and keeps it fresh and invigorating; sweetens foul water and makes it wholesome; works all around and within us a constant miracle, yet with such delicacy and quietness that we never perceive or think of it until we see it with the eye of science.

Compounds.—*Nitric Acid,* HNO_3.*—**Sources.**—This acid is found in nature, combined with Na or K. It is formed in small quantities in the atmosphere by the

Fig. 9.

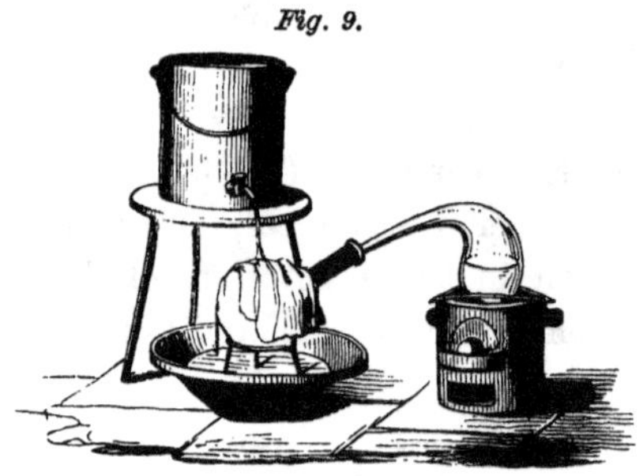

Preparing HNO_3.

union of its elements during the passage of electricity, as in a thunder-storm, and being washed to the earth by rain, is absorbed by the roots of plants.

Preparation.—It is liberated by adding a stronger acid to one of the nitrates. Thus if sulphuric acid (H_2SO_4) and sodium nitrate be heated together in a retort, the

* The molecule of nitric anhydride is N_2O_5; adding the elements of water, we have $N_2O_5 + H_2O = 2(HNO_3)$, or 2 molecules of nitric hydride (nitric acid). In its compounds a metal takes the place of the H; thus, KNO_3, $NaNO_3$, etc.

salt will be decomposed and the acid can be collected in a receiver, cooled by dripping water.

Properties.—It is an intensely corrosive, poisonous liquid.* When pure, it is colorless; but as sold, it has commonly a golden tint from the presence of a lower oxide of N, produced by the decomposing action of the light. It has been obtained in the form of brilliant transparent crystals (nitric anhydride), but they decompose spontaneously. In strength it is next to H_2SO_4. It was formerly called *aqua-fortis,* or strong water. It stains wood, the skin, etc., a bright yellow. It gives up its O readily, and hence is a powerful oxidizing agent.†

Uses.—HNO_3 is employed in dyeing woollen yellow, and in surgery for cauterizing the flesh. It dissolves most of the metals, and in combination with HCl forms *aqua-regia,* the usual solvent of Au. It etches the lines in copperplate engraving, and the beautiful designs on the blades of razors, swords, etc. The process is very simple. The surface is covered with a varnish impervious to the acid, and the desired figure is then sketched in the varnish with a needle. The HNO_3 being poured on, oxidizes the metal in the delicate lines thus laid bare.

* This fact shows the power of chemical affinity. The bland mixture we inhale at every breath is changed to a corrosive poison.

† The following experiments illustrate this property of HNO_3:

1. Mix equal parts of strong HNO_3 and H_2SO_4. Place a little oil of turpentine in a cup out-of-doors, and pour the mixture upon it at arm's length. The turpentine will burn with almost explosive violence.

2. Pour very dilute HNO_3 upon bits of tin. Dense, red fumes (NO_2, hyponitric anhydride) will pass off, and the Sn will be converted into a white oxide, which furnishes what is termed putty powder.

3. Throw crystals of any nitrate on red-hot coals. They will deflagrate on account of the O which they give up to the fire.

4. Soak a strip of blotting-paper in a solution of nitre. It will form "touch-paper," and when lighted will only smoulder.

Nitrous Oxide, N_2O.—**Preparation.**—This gas is made by heating ammonium nitrate (H_4N,NO_3), which decomposes into H_2O and N_2O. (See p. 135.)

Fig. 10.

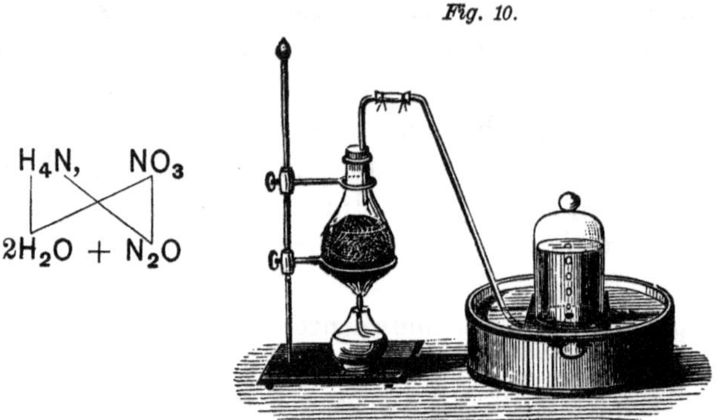

Preparing N_2O.

Properties.—N_2O is a colorless, transparent gas with a faintly sweetish taste and smell. It supports combustion nearly as well as O, and many of the experiments ordinarily performed with that gas will be equally brilliant with N_2O. If breathed for a short time, it produces a peculiar kind of intoxication, often attended with uncontrollable laughter, and hence it has received the popular name of *laughing gas.* The effect soon passes off. If taken for a longer time, it causes insensibility, and is therefore valuable as an anæsthetic in surgical and dental operations.

Nitric Oxide, NO.—**Preparation.**—This gas may be prepared by the action of dilute HNO_3 on copper clippings. The flask (*a*, Fig. 11) will soon be filled with red fumes, but a colorless gas will collect in the jar over water. At the conclusion of the process, the flask will contain a deep blue solution of copper nitrate ($Cu2NO_3$). By filtering and evaporating, the beautiful crystals of this salt may be obtained.

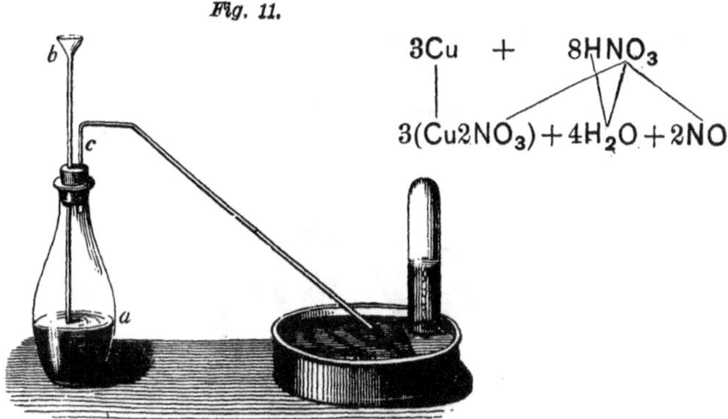

Preparing NO.

Properties.—NO is a colorless, irrespirable gas with a disagreeable odor. Its remarkable property is its affinity for O. Let a bubble escape into the air, and red fumes of hyponitric anhydride (NO_2) will be formed.*

Ammonia, H_3N.—**Source.**—This gas was formerly called hartshorn, because in England it was made from the horns of the hart. It received the name *ammonia,* by which it is now more generally known, from the temple of Jupiter Ammon, near which sal-ammoniac, one of its compounds, was once manufactured. The *aqua-*

* This may be illustrated still more prettily by the following experiment:—Fill a small jar with water colored blue by litmus solution, and pass up into it sufficient NO to occupy about one-third of the bottle; the litmus will not change in color. Now allow a few bubbles of O to rise into the NO; deep red fumes will be formed, which will quickly dissolve, and the blue solution become red. If both the O and the NO be pure, it is possible, by cautiously adding O, to cause a complete absorption of both gases. If common air were used instead of O, only N would then remain in the jar.

ammonia of the shops, which is merely a strong solution of the gas in H_2O, is obtained from the incidental products of the gas-works in large quantities. (See p. 83.) Its pungent odor can often be detected near decaying vegetable and animal matter.

Preparation.—H_3N is ordinarily prepared by heating sal-ammoniac with lime.* The stronger base unites with

Fig. 12.

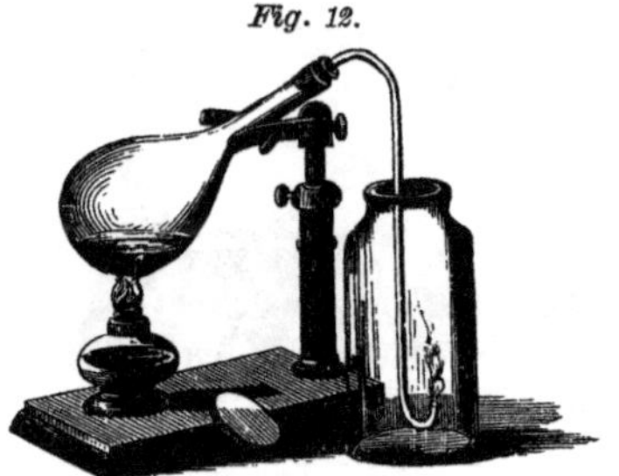

H_3N *burning in* O.

the Cl, and sets the ammonia free. The reaction may be shown as follows:

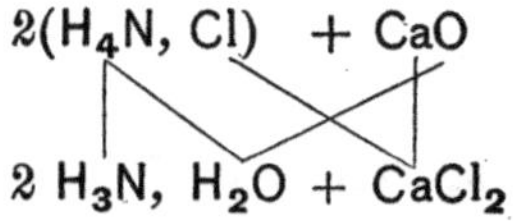

$$2(H_4N, Cl) + CaO = 2\,H_3N, H_2O + CaCl_2$$

Properties.—Water at 60° F. will absorb 700 times its own bulk of the gas.† This solution will produce a blister, and should, therefore, be very much weakened

* This may be illustrated by simply mixing in a cup some powdered sal-ammoniac and lime, when the ammonia may be detected by its odor, and the bluing of moist red litmus-paper.

† Heat a little aqua ammonia in a Florence flask. Dry the vapor and collect in an inverted bottle, to which is fitted a cork and tube, with the inner extremity drawn to a fine point over the spirit-lamp. Insert the cork, and then plunge the bottle into a vessel of water. The water which passes in first will absorb the gas so quickly as to make a partial vacuum, into which the water will rush so violently as to produce a miniature fountain.

before being tasted or touched. It is a strong alkali, and turns the vegetable blues to green; but owing to its volatility this change of color is only temporary. It is, therefore, sometimes termed "the volatile alkali." It neutralizes the most powerful acids, and forms important salts. Its vapor burns in O with a green flame. (See Fig. 12.) Its test is HCl.—*Example:* If we bring a stopple wet with HCl near this gas, it will instantly reveal itself by a dense cloud of white fumes, ammonium chloride (sal-ammoniac), which floats in the air like smoke. The antidote of H_3N is vinegar. Gaseous ammonia becomes liquid at a cold of $-40°$.

Nascent State.—If N and H, the elements of H_3N, be mixed in a receiver, they will not unite chemically, owing to the negative character of N. When, however, any substance is decomposed which contains both of them, as bituminous coal, flesh, etc., at the very instant of their separation they will combine and form H_3N. When elements are thus in the act of leaving their compounds, they are said to be in their "nascent state."

PRACTICAL QUESTIONS.

1. How could you detect any free O in a jar of N?
2. How would you remove the product of the test?
3. In the experiment shown in Fig. 11, why is the gas red in the flask, but colorless when it bubbles up into the jar?
4. How much H_3N can be obtained from 3 lbs. of sal-ammoniac?
5. How much H_2O will be formed in the process?
6. How much CaO will be needed?
7. In separating N, how much air will be needed to furnish a gallon of the gas?
8. How much N_2O can be made from 1 lb. of ammonium nitrate?
9. How much nitric acid can be formed from 50 lbs. of sodium nitrate ($NaNO_3$)?

10. What causes flesh to decompose so much more easily than wood?

11. If a tuft of hair be heated in a test tube, the liquid formed will turn red litmus-paper blue. Explain.

12. Why should care be used in opening a bottle of strong H_3N in a warm room?

13. What weight of N is there in 10 lbs. of HNO_3?

14. How much sal-ammoniac would be required to make 2 lbs. of H_3N?

15. Give illustrations of the replacement of the H in an acid by a metal.

16. What is the difference between liquid ammonia and liquor ammoniæ?

HYDROGEN.

Symbol, H ...Atomic Weight, 1....Specific Gravity, .069.

HYDROGEN means literally a generator of water.

Preparation.—It is always obtained by the decomposi-

Fig. 13.

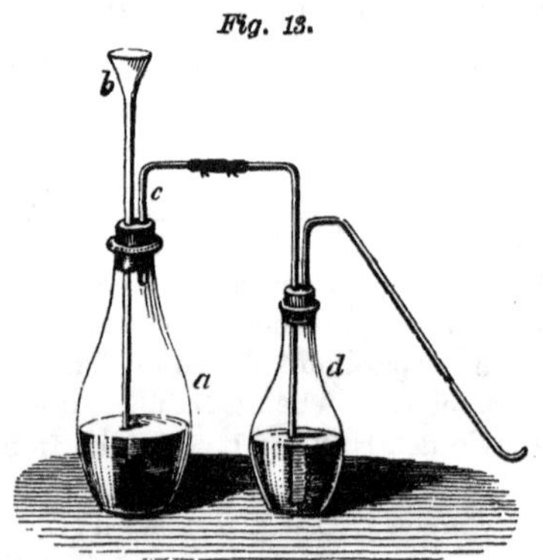

Preparing hydrogen.

tion of H_2O, of which it forms $\frac{1}{9}$ part by weight. If we place in an evolution flask (a common junk bottle will answer) bits of Zn, and then pour through the funnel tube (*b*) H_2O and H_2SO_4, the gas will be evolved abundantly. (See Appendix.) The reaction is as follows:

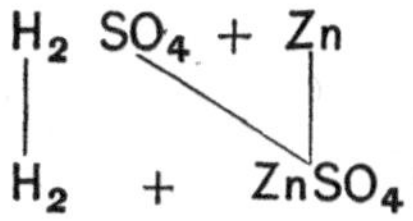

$$H_2\,SO_4 + Zn = H_2 + ZnSO_4$$

The Zn simply takes the place of the H_2. The black specks floating in the liquid are charcoal from the zinc. The milky look is given by the zinc sulphate (white vitriol) which is formed. By evaporating the water, the crystals of this salt may be obtained.

Properties.—H prepared in this manner has a disagreeable odor, from various impurities in the materials used. When pure, it is, like O, colorless, transparent, and odorless. H has the greatest diffusive power of any element; and in attempts made to liquefy the gas, it leaked through the pores of the thick iron cylinders in which it was compressed. It is the lightest of all bodies, being only $\frac{1}{14}$ as heavy as common air. It is not poisonous, although, like N, it will destroy life or combustion by shutting out the life-sustainer, O. When inhaled, it gives the voice a ludicrously shrill tone. It can be breathed for a few moments with impunity, if it be first passed through lime-water to purify it. (See Fig. 13.) Owing to its lightness, it passes out of the lungs again directly. Its levity suggested its use for filling balloons,*

* We read in accounts of fêtes at Paris, of balloons ingeniously made to represent various animals, so that aerial hunts are devised. The animals, however, persistently insist upon ascending with their feet up—a circumstance productive of great mirth in the crowd of spectators.

and it has been employed for that purpose; but coal gas, which contains much H, and is cheaper, is now preferred.

Combustion of H.—A lighted candle, plunged into an inverted jar of H, is extinguished, while the gas itself takes fire, and burns with a feeble flame. One atom of the O of the air unites with two atoms of the H, and the product of the combustion is H_2O, which may be condensed on a cold tumbler, held over a jet of the burning gas. (See Fig. 16.) The Philosopher's Lamp (see Fig. 15) is a more simple means of illustrating the properties of H.

Fig. 14.

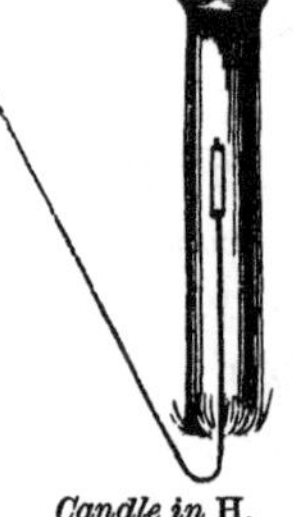

Candle in H.

Fig. 15

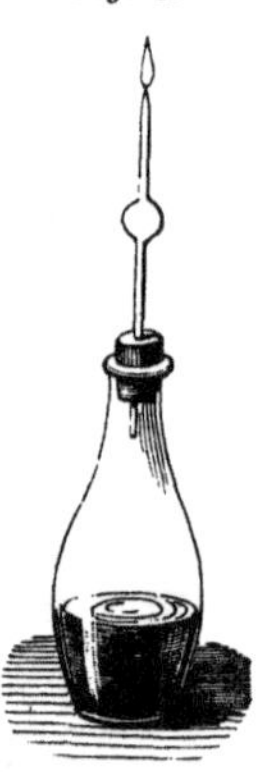

The Philosopher's Lamp.

Mixed Gases.—A mixture of two parts, by measure, of H, with one part of O, or five parts of common air, when ignited, will explode violently.* The heat generated by the union of H_2 and O, expands into steam the drop of H_2O thus formed. Immediately after, the steam being condensed, a vacuum is produced and the particles of air rushing in to fill the empty space, by their collision against each other, cause the deafening sound. While the detonation is so great, the force is slight, as may be shown by exploding, in the hand, soap-bubbles blown with the gases. H

* The H gun—which is simply a tin tube, closed at one end, and provided with a cork at the other, having a priming-hole at the side—is used to illustrate this fact. It may be filled over the Philosopher's Lamp when that is not ignited. The gas is allowed to pass in until the gun is about a fifth full, as nearly as one can guess, when the gun is removed and the gases ignited at the priming-hole.

Fig. 16.

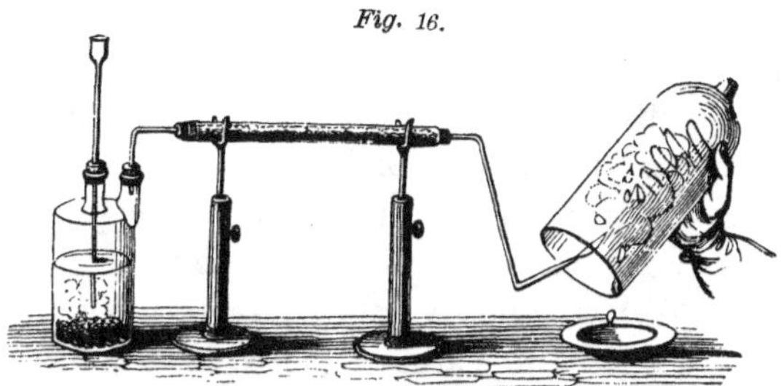

H_2O *formed by burning* H.

and O may be mingled in the right proportion for combustion, and though kept for years, there will be no change.

Fig. 17.

Transferring gases.

The different atoms lie against one another quietly, with no manifestation of their chemical affinity, until suddenly, at the contact of the merest spark of fire, they rush together with a crash like thunder, and uniting, form the bland, passive liquid—water.

Action of Spongy Platinum.—A piece of spongy platinum placed in a jet of H will ignite it. This curious effect seems to be produced in the following way: The atoms of H and the O of the air are brought so closely together in its minute pores that they unite, and the heat thus generated sets fire to the gas. This action is nicely shown by the instrument represented in Fig. 18. It was formerly used by chemists as a convenient way of obtaining a light in the laboratory. Friction matches have superseded this ingenious invention.

Fig. 18.

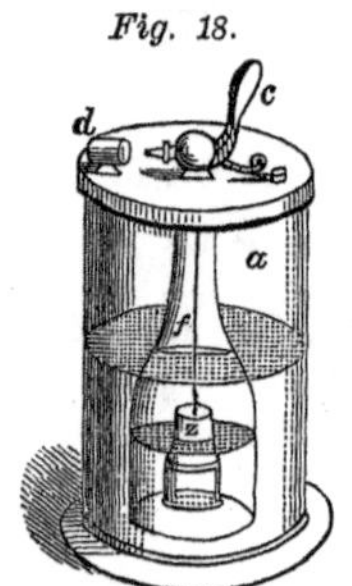

*Döbereiner's Lamp.**

Heat of Burning H.—A hydrogen flame gives little light, but great heat. In H and O, existing as gases, there is stored a vast amount of latent energy. (*Physics*, pp. 36, 185.) When they unite by chemical affinity, this force is set free. "In the union of 16 lbs. of O and two of H, sufficient potential force is developed to raise 40,000,000 lbs. a foot high."

Hydrogen Tones.†—A singular illustration of the

* Z is a piece of zinc suspended in a mixture of H_2SO_4 and H_2O. At the top is a stop-cock, by turning which the gas is allowed to pass out from the receiver *f*. It strikes upon a piece of spongy platinum, and ignites with a slight explosion.

† Another illustration of singing hydrogen may be represented in the following manner: Make a jar of heavy tin, in the form of a double cone, 12 inches long and 4 inches in diameter. At one apex fit a nozzle and cork; at the other,

laws of sound can be given by simply holding a long glass tube, by means of a suitable clamp, over a minute jet of burning H. At first no effect will be produced; but as we slowly introduce the jet further and further into the tube, a faint sound is heard, apparently in the far-off distance. It gradually approaches, and finally bursts into a shrill, continuous, musical note — the key-note of the heated column of air within the tube. The flame is momentarily extinguished and re-lighted with a slight explosion, the rapid repetition of which is supposed to produce

Fig. 19.

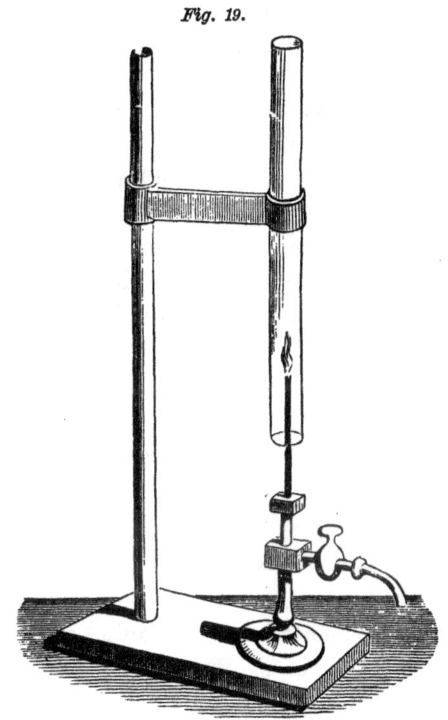

Hydrogen tones.

make several minute openings. Cover the holes with sealing-wax, and draw the cork; then fill the jar with H, and replace the cork. When ready for use, hold the jar in a vertical position, remove the wax from at least one orifice, ignite the H at that point, and draw the cork. Still hold the jar quietly, and in a minute or two the tiny jet of H will begin to sing like a swarm of mosquitoes, buzzing and humming in a most aggravating way until, unexpectedly, the scientific music ends in a loud explosion.

the musical note. Indeed, the explosions may be made so slow that the quivering of the flame can be seen, and the sound cease to be continuous as before. Let us now place the tube at a point where no clapping of hands or unusual sound will start it into song. Let various tones be produced from a violin, and we shall find the flame responding only to that tone which is the key-note of the tube, or its octave. The violin player will have perfect control of this scientific music, and can start, stop, or throw it into violent convulsions, even across a large hall. Tubes of different sizes and lengths will give tones of diverse character and pitch.* The waves of sound from the instrument augmenting or interfering with those in the tube probably produce these phenomena. (See *Physics*, p. 141.)

WATER.

THE COMPOSITION of H_2O is proved by analysis and synthesis—*i. e.*, by separating the compound into its elements, and by combining the elements to produce the compound. We can analyze it in the manner already shown in preparing H, or by passing through it a galvanic current, when the O will appear in bubbles of gas at the positive pole, and the H in a similar way at the negative. In the synthetic method, we mix the two gases, and unite them as we have before by an electric spark. The blacksmith decomposes water when he sprinkles it on the hot coals in his forge. The H burns with a

* The singing of the hydrogen flame may be illustrated more simply by holding the beaks of broken retorts, or large tubes of any kind, over the flame of the Philosopher's Lamp. Jets of different sizes may be made by drawing out glass tubing over the spirit-lamp.

pale flame, while the O increases the combustion. Thus, in a fire, if the engines throw on too little water, it may be decomposed, and add to the fury of the flame.* To "set the North River on fire" is only a poetical exaggeration.

Fig. 20.

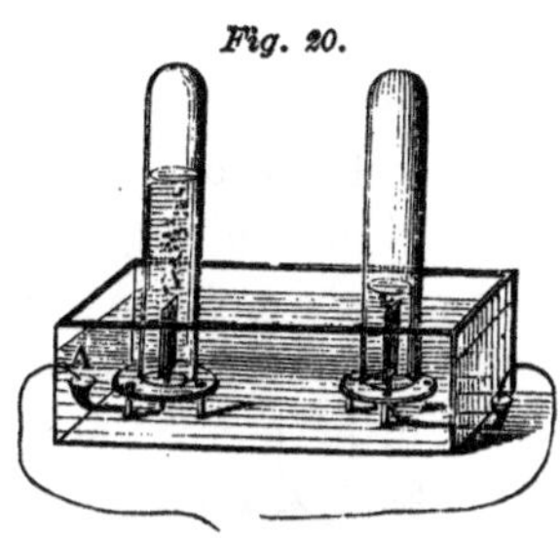

Analysis of water.

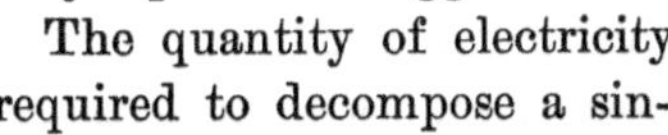

The quantity of electricity required to decompose a single grain of water is estimated to be equal to a powerful flash of lightning. The enormous force necessary to tear these two elements from each other shows the wonderful strength of chemical attraction.† We thus see, that in a tiny drop of dew there slumbers the latent power of a thunderbolt.

Water in the Animal World.—The abundance of water very forcibly attracts the attention. It composes perhaps $\frac{4}{5}$ of our flesh and blood. Man has been facetiously described as 12 lbs. of solid matter wet up in six pails of water. All plumpness of flesh, and fairness of the cheek, are given by the juices of the system. A few ounces of water and a little charcoal constitute the principal chemical difference between the round, rosy face of sixteen,

* "No more heat is produced by the action of the H_2O, but it is in a more available form for communicating heat. The steam in contact with incandescent charcoal is decomposed—the O going to the C to form CO_2, and the H being set free. If the C is abundant, and the heat high, the CO_2 is also decomposed, and double its volume of CO formed. The inflammable gases, H and CO, mingled with the hydrocarbons always produced, are ignited, making the billows of flame which sweep over a burning building."—S. P. SHARPLES.

† The force needed to separate them becomes latent in the gases as a potential force, and when they are burned at any time will be set free as sensible heat—a dynamic force.

and the wrinkled, withered features of three-score and ten. To supply the constant demand of the system for water, each adult, in active exercise, needs about three pints per day, or over half a ton annually. (See *Physiology*, p. 220.) When we pass to lower orders of animals, we find this liquid still more abundant. Sunfishes are little more than organized water. Professor Agassiz analyzed one found off the coast of Massachusetts, which weighed 30 lbs., and obtained only half an ounce of dried flesh. Indeed, naturalists state that an entire class of animals (hydrozoa), to which belong the jelly-fish, medusa, etc., is composed of only ten parts in a thousand of solid matter. (See *Zoology*, p. 269.)

Water in the Vegetable World.—In the vegetable world we find it abundant. Wood is composed of 6 parts charcoal and 5 parts water, with a little mineral matter comprising the ashes. Bread is half water; and of the potatoes and turnips cooked for our dinner, it comprises 75 parts of one and 90 of the other. The following table shows the proportion in common vegetables, fruits, and meats:

Mutton....	.71	Trout.....	.81	Cabbage......	.92
Beef.......	.74	Apples....	.80	Cucumbers...	.97
Veal.......	.75	Carrots....	.83	Watermelons	.98
Pork.......	.76	Beets.....	.88		

Water in the Mineral World.—Bodies in which the water is chemically combined in definite proportions, are often called *hydrates*. In the image which the Italian pedler carries through our streets for sale, there is nearly 1 lb. of H_2O to every 4 lbs. of plaster of Paris. One-third of the weight of any ordinary soil is this same liquid. Each pound of strong nitric acid contains 2$\frac{2}{7}$ oz. of water, which, if removed, would destroy the acid itself. If we expel the water from oil of vitriol, it will lose its acid properties, and we can handle it with impunity. In bodies which are capable of crystallizing, it seems to determine the form and general appearance, and is called "the water of crystallization." If we evaporate this from blue vitriol, it will lose its color and become white like flour.* A few drops of H_2O will restore the blue. If we expel this from alum, it will puff up, and the transparent crystals will dry into an incoherent mass. Many salts *effloresce, i. e.*, part with their water of crystallization on exposure to the air, and crumble into a white powder.

Water as a Solvent.—Water, having no taste, color, or odor itself, is perfectly adapted to be the universal solvent. It becomes at pleasure sweet, sour, salt, bitter, nauseous, and even poisonous. Had water any taste, the whole science of cookery would be changed, since each substance would partake of the one universal watery flavor.

Pure Water.—Rain-water, caught after the air is thoroughly cleansed by previous showers, and at a distance from the smoke of cities, is the purest natural water known. It is tasteless, yet its insipidity makes it seem to us very ill-flavored indeed. We have become so accustomed to the taste of the impurities in hard water, that they have become to us tests of its sweetness and pleasantness.

River Water, though it may have less mineral mat-

* This may be easily shown by filling the bowl of a tobacco-pipe with crystals of the salt, and heating them over a lamp or in the fire until the water of crystallization is expelled. Alum may be made anhydrous in the same way.

ter than spring water, is often unfitted for drinking on account of the organic matter it contains. Happily, running water has in itself a certain purifying power, owing to the air which it holds in solution; so that, paradoxical as it may seem, organic substances are burned in it as certainly as they would be in a stove. Still, in order to avoid any danger, river water should be filtered through charcoal or sand before using.*

Hard Water.—As water percolates through the soil into our wells, it dissolves the various mineral matters characteristic of the locality. The most abundant of these are lime,† salt, and magnesia. The former produces a *fur* or coating on the bottom of our tea-kettles, if we live in a limestone region. When we put soap in such water, it curdles—*i. e.*, it unites with the lime (CaO), forming a new, or lime soap, which is insoluble in H_2O. H_2O containing an excess of mineral matter is unwholesome; yet it is probable that the sparkling hard waters of the limestone districts are relished, not only because they are pleasant to the eye and agreeable to the taste, but on account of some hygienic properties in the excess of CO_2 they contain, and possibly because the CaO acts medicinally on the system.‡

* A weak solution of potassium permanganate is an excellent test of the presence of organic matter. Place the water to be examined in a glass, and add a little permanganate; if organic matter is present the violet permanganate solution is decolorized as fast as added until all the organic matter is oxidized.

† It is a fact worthy of note that lime and oxide of iron, which are frequently found in H_2O, the latter generally in minute quantities, are both healthful; while the oxides of the other metals are poisonous. Were zinc or barium, for instance, as common near our homes as iron or calcium, wholesome drinking water would be rarely, if ever, found. By a wise arrangement of an ever-watchful Providence, those dangerous metals are rare, and hidden far from the haunts of man.

‡ The French authorities are so well satisfied of the superiority of hard water, that they pass by that of the sandy plains, near Paris, and go far away to the chalk hills of Champagne, where they find water even harder than that of Lon-

Sea-Water.—The most abundant mineral in the ocean is common salt. Yet it contains traces of every substance soluble in water, which has been washed into the sea from the surface of the continents during all the ages of the past. Its saline constituents are now in the proportion of about ½ oz. to 1 lb. This amount may be slowly increasing, as the water which evaporates from the surface is comparatively pure, containing only a mere trace of a few substances, which give to the sea-breeze its peculiar bracing, tonic influence. In this way, the water of the Salt Lake has become a strong brine, nearly ⅓ of its whole weight consisting of saline matter. This condition would soon disappear if an outlet could be provided.

Water Atmosphere.—As the world of waters is inhabited, it also has its atmosphere.* Inasmuch as the H_2O dilutes the O in part, it does not need so much N as the common air. It is accordingly composed of over ⅓ O instead of only ⅕. The air so rich in O thus absorbed by the water gives to it life and briskness. If it be expelled by boiling, the water tastes flat and insipid.

Paradoxes of Water.—"Cold contracts," is the law of physics; but as H_2O cools, it obeys this principle only as far as 39° F. Then it slowly expands, cooling down to 32°, its freezing point, when its crystals suddenly dart out at angles to each other, and thus, increas-

don; giving as a reason for the preference that more of the conscripts from the soft-water districts are rejected on account of the want of strength of muscle, than from the hard-water districts; from which they conclude that the calcareous matter is favorable to the formation of the tissues.

* Fish inhale O through the fine silky filaments of their gills. When a fish is drawn out of H_2O, these dry up, and it is unable to breathe, although it is in a more plentiful atmosphere than it is accustomed to enjoy.

ing in size about $\frac{1}{12}$, it congeals to ice. By this wise arrangement, ice is lighter than water, and so swims on top; otherwise our rivers would freeze solid, killing the fish and aquatic plants. The longest summer could not melt such an immense mass of ice. But now the blanket that Nature kindly weaves over the rivers and ponds keeps their finny inhabitants warm and comfortable till spring; then she floats it south to melt under a hotter sun. We give to water such contradictory terms as "hard" and "soft," "fresh" and "salt." H_2O seems the most yielding of substances, yet the swimmer who falls on his face, instead of striking head foremost, appreciates the mistake, and we could drive a nail into a solid cube of steel as easily as into a hollow one perfectly filled with H_2O. H is the lightest substance known, and O is an invisible gas; yet they unite and form a liquid whose weight we have often experienced, and a solid which makes a pavement hard like granite. H burns readily, and, when mixed with O, explodes most fearfully; O supports combustion brilliantly—yet the two combined are used to extinguish fires. H or O in excess would destroy life; H_2O is so essential to it that thirst causes a lingering, painful death.

Uses of Water.—The uses of H_2O are as diverse as they are practical. Its properties fit it for a wonderful variety of operations in nature. Its office is not merely to moisten our lips on a hot day, to make a cup of coffee, to lay the dust in the street, and to sprinkle our gardens; it has grander and more profound uses than any of these. *Water is the common carrier of creation.* It dissolves the elements of the soil, and, climbing as sap up through the delicate capillary tubes of the plant, furnishes the leaf with the materials of its growth. It flows through the body as blood, floating to every part of the system the life-sustaining O, and the food necessary for repairs and for building up the various parts of the "house we live in." It comes in the clouds as rain, bringing to us the heat of the tropics, and tempering our northern climate, while in spring it floats the ice of our rivers and lakes away to warmer seas to be melted. It washes down the mountain side, levelling its lofty summit and bearing mineral matter to fertilize the valley beneath. It propels water-wheels working forges and mills, and thus becomes the grand motive-power of the arts and manufactures. It flows to the sea, bearing on its bosom ships conducting the commerce of the world. It passes through the arid sands, and the desert forthwith buds and blossoms as the rose. It limits the bounds of fertility, decides the founding of cities, and directs the flow of trade and wealth.

PRACTICAL QUESTIONS.

1. Why, in filling the hydrogen gun, do we use 5 parts of common air to 2 of H, and only 1 part of O to 2 of H ?
2. Why are coal cinders often moistened with H_2O before using ?
3. What injury may be done by throwing a small quantity of H_2O on a fire ?
4. Why does the hardness of water vary in different localities ?
5. What causes the variety of minerals in the ocean? Is the quantity increasing ?
6. Is there not a compensation in the sea-plants, fish, etc., which are washed back on the land ?
7. Since "all the rivers flow to the sea," why is it not full ?
8. What is the cause of the tonic influence of the sea breeze ?
9. When fish are taken out of the water, and thus brought into a more abundant atmosphere, why do they die ?

10. Do all fish die when brought on land?

11. What weight of water is there in a cwt. of sodium sulphate (Na_2SO_4, $10H_2O$), or Glauber's salt?

12. What weight of water in a ton of alum ($KAl2SO_4$, $12H_2O$)?

13. How much water would it require to change 5 lbs. of nitric anhydride to nitric acid?

14. How does the air purify running water?

15. What is the action of potassium permanganate as a disinfectant?

16. Why does lime sometimes soften hard water when added to it?

17. What weight of H can be obtained from a gallon of water?

18. In decomposing H_2O, 65 parts by weight of Zn yield 2 parts by weight of H. How much Zn must be employed to obtain 100 lbs. of H?

19. How much $KClO_3$ would be required to evolve sufficient O to burn the H produced by the decomposition of 2 lbs. of H_2O?

20. How much O would be required to oxidize the metallic Cu which could be reduced from its oxide by passing over it, when white-hot, 20 gr. of H gas?

21. How much O would be required to oxidize the metallic Fe which could be reduced in the same manner by 10 grs. of H gas?

22. Why are rose-balloons so buoyant?

23. How much H must be burned to produce a ton of water?

CARBON.

Symbol, C. Atomic Weight, 12. Specific Gravity of Diamond, 3.3 to 3.5.

Source.—C is one of the most abundant substances in nature, forming nearly one-half of the entire vegetable kingdom, and being a prominent constituent of limestone, corals, marble, magnesian rocks, etc. We find it in three distinct forms or allotropic conditions—viz., the *diamond*, *graphite*, and *amorphous carbon*. This last term means without crystalline form, and includes gas-carbon, charcoal, lamp-black, coal, coke, peat, soot, bone-black and ivory-black. In each of these various substances C possesses different properties; yet any impurities it may contain seem entirely incidental, and not at all necessary to its new state.

Proof of this Allotropic state.—Chemists have changed most of these substances into other allotropic forms. Thus, common charcoal has been turned into graphite, mineral coal into gas-carbon, the diamond into coke. All of them, when heated in the open air, unite with the same quantity of O, forming precisely the same compound—carbonic anhydride—from which the C can be obtained again in the form of charcoal.

The Diamond is *pure carbon* crystallized. It is the hardest of all known substances, scratches all other minerals and gems, and can be cut only by its own dust. It is infusible, but will burn at a high temperature. It is found in various parts of the world—North Carolina, Georgia, Borneo, Brazil, and South Africa. In 1858, Brazil furnished 120,000 carats.* They usually appear as semi-transparent, rounded pebbles, enclosed in a thin, brownish, opaque crust, which being broken reveals the brilliant gem within. They are of various tints, though often colorless and perfectly transparent. The last are most highly esteemed, and, from their resemblance to a drop of clear spring-water, are called diamonds of the "first water." They are exceedingly brittle, and valuable gems are said to have been broken by simply

* A carat is equal to 4 grs. Troy. The term is derived from the name of a bean which, when dried, was formerly used in weighing by the diamond merchants in India.

falling to the floor. Nothing definite is known concerning the origin of this gem.*

The Diamond is Ground by means of its own powder. Being fitted to the end of a stick or handle, it is pressed down firmly against the face of a rapidly revolving wheel, covered with diamond-dust and oil. This, by its friction, removes the exposed edge and forms a *facet* of the gem. There are three forms of cutting—the *brilliant*, the *rose*, and the *table*.

Fig. 21.

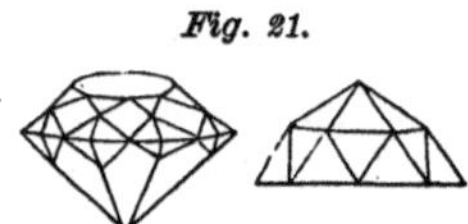

The brilliant. *The rose.*

The brilliant has a flat surface on the top, with facets at the side, and also below, the latter terminating in a point, so arranged as to refract the light most brilliantly. This form shows the gem to the best advantage, but is used only in large, thick stones, as it sacrifices nearly half the weight in cutting. The rose is flat beneath, while the upper surface is ground into triangular facets, terminating at a common vertex. The table form is employed for thin specimens, which are merely ornamented by small facets on the edge. The diamond is valued not

* Although the diamond is simply pure carbon, yet it has never been made by any chemical process. Minute diamonds, it is said, have been separated from carbon compounds by long-continued voltaic action, but they were invisible except with a microscope. The value of the diamond varies with the market; the *general* rule is as follows: a gem ready for setting, of one carat weight, is worth $150 to $180; beyond this size, the estimated value increases according to the square of the weight, but in case of large stones is generally much less than that amount, although rare beauty or size may greatly enhance the price. The *Kohinoor* (mountain of light, now among the crown jewels of England) weighs 103 carats, yet is valued at $10,000,000. Owing to the discovery of many large diamonds in South Africa, the value of such stones has much decreased of late. The smaller ones, however, are becoming more expensive on account of the greater demand for them. The South African diamonds are seldom colorless, having generally a yellowish tint. Paste diamonds are now made in Paris, which are so perfect an imitation that only experts can distinguish them from the real gems.

alone for its rarity and high refractive power, by which it flashes such vivid and brilliant colors, but also for its mechanical uses. For cutting glass, the curved edges of the natural crystal are used.

Graphite or *Plumbago* is also called black-lead, because on paper it makes a shining mark like lead. It is found at Ticonderoga, N. Y., Brandon, Vt., and Sturbridge, Mass. It is supposed to be of vegetable origin.

Uses.—It is chiefly useful in pencils. For this purpose a mixture of black-lead, antimony, and sulphur—the proportion of these ingredients determining the hardness of the pencil—is melted and cast into blocks, which are then sawed into thin slips, as seen in common pencils.* Though graphite seems very soft, yet its particles are extremely hard, and the saws used in cutting it soon wear out. We notice this property in sharpening a pencil with a knife. Graphite mixed with clay is made into black-lead crucibles. These are the most *refractory* known, and are used for melting gold and silver. It is also sold as "British lustre," "carburet of iron," "stove polish," etc., which are employed for blacking stoves and protecting iron from rusting.

Gas-Carbon is formed on the interior of the retorts used in coal-gas works. It has a metallic lustre, and will scratch glass.

Charcoal is made by burning piles of wood, so covered over with turf as to prevent free access of air. The volatile gases, water, etc., are driven off, and the C

* For drawing-pencils, pure graphite powder is subjected to such enormous pressure that the particles are brought near enough together for the attraction of cohesion to hold them in a solid form, when the pressure is removed. This block is then sawed into prisms, which are fitted into cylinders of cedar-wood.

left behind. This forms about $\frac{3}{4}$ of the bulk of the wood and $\frac{1}{4}$ its weight. Charcoal for gunpowder and for medi-

Fig. 22.

Making charcoal.

cinal purposes is prepared by heating willow or poplar wood in iron retorts.

Properties.—It is the most unchangeable of all the elements, so that even in the charcoal we can trace all the delicate structure of the plant from which it was made. It is insoluble in any ordinary liquid. None of the acids, except nitric, corrodes it. No alkali will eat it. Neither air nor moisture affects it. Wheat has been found in the ruins of Herculaneum that was charred 1800 years ago, and yet the kernels are as perfect as if grown last harvest. The ground ends of posts are rendered durable by charring. Indeed, some were dug up not long since in the bed of the Thames which were placed there by the ancient Britons to oppose the passage of Julius Cæsar and his army. A cubic inch of fine charcoal has, it is said, 100 feet of surface, so full is it of minute pores. These absorb gases by capillary attraction to an almost incredible extent. A bit of C will take up 90 times its bulk of ammonia. As the various gases and the O of the air are brought so closely together within its pores, rapid oxidation is produced, as in the case of spongy platinum (see p. 54). Pans of charcoal soon purify, and sweeten the offensive air of a hospital. Foul water filtered through C loses its impurities. Beer by this process parts not only with its color but with its bitter taste. Ink is robbed of its value and comes out clear and transparent as water.

Deoxidizing or Reducing Action of C.—At a high temperature the attraction of C for O is powerful. In the heat of a furnace it will take it from almost the stablest compounds. This fact gives to charcoal great value in the arts. Nearly all the metals and many of the other elements are locked up in the rocks with O, and C is the key made by the Creator for unlocking the treasure-houses of nature for the supply of our wants. By noticing the process of preparing zinc, iron, phosphorus, etc., we shall see the importance of this property of C. A very pretty illustration is shown by placing a few grains of litharge (PbO) on a flat piece of charcoal, and directing upon it the flame of a blowpipe. The metal will immediately appear in little sparkling globules.

Fig. 23.

PbO *on charcoal.*

Soot is unburnt carbon which passes off from a lamp or fire when there is not enough O present to combine with all the C of the fuel. This, therefore, comes away in flakes, and blackens the chimney of the lamp, or lodges in the chimney of the house. After a time, a large quantity having collected, we are startled by the cry, "The chimney is on fire!" while with a great roar and flame the soot burns out. This unpleasant occurrence is much more frequent when green wood is used for fuel. The H_2O of the wood absorbs much of the heat of the fire, and so permits the C to pass off unconsumed.

Lampblack is obtained by imperfectly burning pitch or tar. The dense cloud of smoke is conducted into a chamber lined with sacking, upon which the soot collects. It is largely used in painting. It is mixed with clay to form black drawing-crayons, and with linseed oil to make printers' ink. Lampblack has peculiar properties which fit it for printing. Nothing in nature could supply its place. No matter how finely it is pulverized, it retains its dead-black color. The minutest particle is as black as the largest mass. It is insoluble in all liquids. It never decays. The paper may moulder; we may even burn it; and still, in the ashes, we can trace the form of the printed letter. The ancients used an ink composed of gum-water and lampblack, and manuscripts have been exhumed from the ruins of Pompeii and Herculaneum which are yet perfectly legible.

Animal Charcoal, or bone-black, is made by burning bones in close vessels. Mixed with H_2SO_4 it forms the basis of paste-blacking. It is largely used by sugar refiners (p. 190). Common vinegar filtered through it becomes the white vinegar of the pickle manufacturers.

Mineral Coal.—This was formed at an early period of the world's history, called the Carboniferous Age. The earth was then pervaded by a genial, tropical climate. The air was denser and richer with vegetable food than now. The earth itself was a swamp, moist and hot, in which simple ferns towered into trunks a foot and a half in diameter; and where plants like those which creep at our feet to-day, or are known only as rushes or grasses, grew to the height of lofty trees. The song of bird or hum of insect rarely echoed through the mighty fern-forests; but a strange and grotesque vegetation flourished with more than tropical luxuriance. In these swamps accumulated a vast deposit of leaves and fallen trunks which, under the water, gradually changed to charcoal. In the process of time, the earth settled at various points, and floods poured in, bringing sand, pebbles, clay, and mud, filling up all the spaces between the trees that were standing, and even the hollow trunks themselves. The pressure of this soil and the internal heat of the earth combined to expel the gases from the vegetable deposits, and convert them into mineral coal.* In time this section was elevated again, and another forest flourished, to be in its turn converted into coal. Each of these alternate elevations and depressions produced a layer of coal or of soil. In these beds of coal we now find the trunks of trees, the outlines of trailing vines, the stems and leaves of plants as perfectly preserved as in a herbarium, so that, to the botanist, the flora of the Carboniferous age is nearly as complete as that of our own.

* Where this process was nearly complete, anthracite coal, and where only partially finished, bituminous coal, was formed. The greater the pressure, the harder and purer the carbon produced; unless, however, the covering was not sufficiently porous to allow the gases to escape, when bituminous coal was the result.

Coke is the refuse of gas-works, obtained by distilling the water, tar, and volatile gases from bituminous coal. It is burned in locomotives, blast-furnaces, etc.

Peat is an accumulation of half decomposed vegetable matter in swampy places.* It is produced mainly by a kind of moss which gradually dies below as it grows above, and thus forms beds of great thickness. Sometimes, however, plants may grow in the form of a turf, and decay, thus collecting a vast amount of vegetable *debris*. This gradually undergoes a change, and becomes a brownish black substance, loose and friable in its texture, resembling coal, but, unlike it, containing 20 to 30 per cent. of O. Peat is used in large quantities as a fuel. For this purpose it is cut out in square blocks and dried in the sun. In some beds it is first finely pulverized, then pressed into a very compact form like brick.

Muck is an impure kind of peat, not so fully carbonized; though the term is frequently applied to any black swampy soil which contains a large quantity of decaying vegetable matter. Like charcoal, it absorbs moisture and gases, and is therefore used as a fertilizer.

Various Forms and Uses of Carbon. — We have seen in what contrary forms C presents itself. It is soft enough for the pencil-sketch, and hard enough for the glazier's use. Black and opaque, it expresses thought on the printed page: clear and brilliant, it gleams and flashes in the diadem of a king. In lampblack, it frequently takes fire spontaneously; in graphite, it resists the heat of the fiercest flame; in the diamond, it is an

* These peat-beds are of vast extent. One-tenth of Ireland is covered by them. A bed near the mouth of the River Loire, is said to be fifty leagues in circumference.

insulator; while in charcoal, it is so perfect a conductor of electricity that it is packed about the foot of lightning-rods to complete the connection with the earth. We burn it in our lamps, and it gives us light; we burn it in our stoves, and it gives us heat; we burn it in our engines, and it gives us power; we burn it in our bodies, and it gives us strength. As fuel, it readily unites with O, yet we spread it as stove-polish on our iron-ware to keep the metal from rusting. It gives firmness to the tree and consistency to our flesh. It is the valuable element of all fuel, burning oils, and gases. Thus it supplies our wants in the most diverse manner, illustrating in every phase the forethought of that Being who fitted up this world as a home for His children. Infinite Wisdom alone would have stored up such supplies of fuel and light, and hidden them far under the earth away from all danger of accidental combustion, or anticipated the requirements alike of luxury and the arts.

Compounds.—*Carbonic Anhydride*, CO_2.—**Source.**—This gas is commonly known as Carbonic Acid. It is found combined with Ca, in a large class of salts, known as the carbonates, viz., limestone, marble, chalk, etc., forming nearly one-half of their weight, and almost one-seventh of the crust of the earth. It comprises $\frac{4}{10,000}$ of the atmosphere. It is produced throughout nature in immense quantities. Wherever C burns, in fires, lights, decay, fermentation, volcanoes—in a word, in all those various forms of combustion of which we spoke under the subject of O, where that gas unites with C, CO_2 is the result. Each adult exhales daily about $8\frac{1}{2}$ oz. of carbon changed to this invisible gas. Each bushel of charcoal, in burning, produces not far from 2500 gallons. A lighted candle gives off about 4 gallons per hour.*

Preparation.—For experimental purposes it is prepared by pouring hydrochloric (muriatic) acid on marble or chalk. The reaction may be represented as follows:

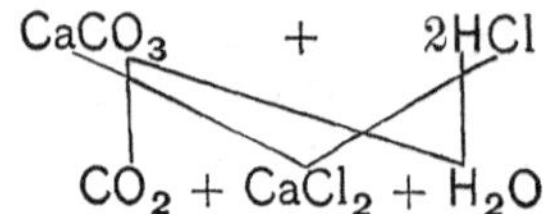

Fig. 24.

Preparing CO_2.†

The CO_2 is liberated rapidly and may be gathered by displacing the air (see Fig. 24), while the calcium chloride remains dissolved in the water of the flask.

* Burn a piece of charcoal or a candle in a jar of O. Pour in a little lime-water and shake it well, when there will be a precipitation of chalk (calcium carbonate). Hold a jar of air over a burning lamp or jet of coal-gas, or breathe into the jar and apply the test.

† Twist a wire around the neck of a small, wide-mouthed vial, to serve as a bucket. Dip the CO_2 with it *upward* from the jar and test with a lighted match. Dip the H (Fig. 14) *downward*, and test in same way. This illustrates in a striking manner the difference between the gases in respect to specific gravity and combustion.

The test of CO_2 is clear lime-water. If we expose a saucer of lime-water to the air, the surface of the solution will soon be covered with a thin pellicle of calcium carbonate (carbonate of lime), thus showing that there is CO_2 in the atmosphere; or if we breathe by means of a tube through lime-water, the solution will become turbid and milky, thus proving the presence of CO_2 in our breath: by breathing through the liquid a little longer it will become clear, as the carbonate will dissolve in an excess of CO_2.

Properties. — CO_2 is a colorless, odorless, transparent gas, with a slightly acid taste, and is a non-supporter of

Fig. 25.

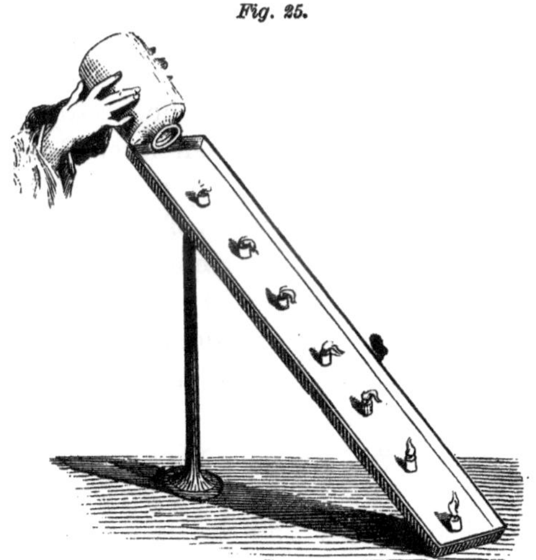

Pouring CO_2 *down an inclined plane.*

combustion. On account of its being heavier than air many amusing experiments can be performed with it. It will run down an inclined plane, can be poured from one dish to another, drawn off by a syphon, dipped up with a bucket like water, or weighed in a pair of scales like lead.

Fig. 26.

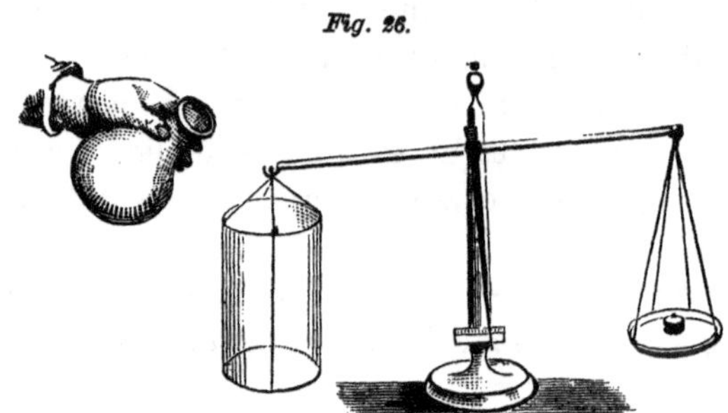

Weighing CO_2 *with a pair of scales.*

To show the C *in* CO_2 hold a strip of Mg foil in a flame until well ignited, then insert in a jar of the gas. White flakes of magnesium oxide* (MgO) mixed with black particles of charcoal will be deposited.

Asphyxia.—CO_2 accumulates in old wells and cellars, where it has cost the lives of many incautious persons.† The test of lowering a lighted candle should always be employed. If that be extinguished, your life would be in danger of "going out" in the same way, should you descend. The gas may be dipped out like water, or the well may be purified by lowering pans of slacked lime, or lighted coals which, when cool, will absorb the noxious gas. The coals may be re-ignited, and

* These may be dissolved by dilute HNO_3, and the black C made more distinct.

† "Three or four per cent. of CO_2 in the air acts as a narcotic poison by preventing the proper action of the air upon the blood."—MILLER.

Fig. 27.

Pouring CO_2 *on a light.*

lowered repeatedly until the result is reached.* Persons have been suffocated by burning charcoal in an open furnace in a closed room.† In France, it is not unusual to commit suicide in this manner. The antidote is to bring the sufferer into the fresh air, and dash cold water upon his face. In the celebrated *Grotto del Cane*, near Naples, the gas accumulates upon the floor, so that a man living near amuses visitors, for a small fee, by leading his dog into the cave. He experiences no ill effects himself, but the dog falls senseless. On being drawn into the open air, the animal soon revives, and is ready to pick up his bit of black bread and enjoy the reward of his scientific experiment.

CO_2 *in Mines.*—Miners call CO_2 *choke-damp.* It is produced by the combustion of *fire-damp* (see p. 81), which accumulates in deep mines,‡ and when mixed with air, burns like gunpowder, forming dense volumes of CO_2, which instantly destroys the lives of all who may have escaped the flames of the explosion.§ CO_2 has been used

* A well, in which a candle would not burn within twenty-six feet of the bottom, was thus purified in a single afternoon.

† The fumes of burning charcoal owe their deadly property largely to the presence of CO (page 81), one per cent. of which in the air causes headache.

‡ The word *gas* was first used in the seventeenth century. Explosions, strange noises, and lurid flames had been seen in mines, caves, etc. The alchemists, whose earthen vessels often exploded with terrific violence, commenced their experiments with prayer, and placed on their crucibles the sign of the cross—hence the name crucible from *crux* (gen. *crucis*), a cross. All these manifestations were supposed to be the work of invisible spirits, to whom the name *gâst* or *geist*, a ghost or spirit, was applied. The miners were in special danger from these unseen adversaries, and it is said that their church service contained the petition, "From spirits, good Lord, deliver us!" The names "spirits of wine," "spirits of nitre," etc., are a relic of the superstitions of that time.

§ Where CO_2 alone is found, it is not considered as dangerous as the *fire-damp*,

for the purpose of extinguishing fires in coal-mines. A mine near Sterling, England, had burned for thirty years, consuming a seam of coal nine feet thick, over an area of twenty-six acres. CO_2, eight million cubic feet of which were required, was poured into the mine, in a continuous stream, day and night, for three weeks. The mine was then cooled with water, and within a month from the commencement of the operation was ready for the resumption of work.

Absorption of CO_2 by Liquids.—Water dissolves its own volume of CO_2 under the ordinary pressure of the atmosphere, forming a solution of carbonic acid; $CO_2 + H_2O$ becoming H_2CO_3. With increased pressure a much greater amount will be absorbed. "Soda water" contains no soda, but is simply H_2O saturated with CO_2 in a copper receiver strong enough to resist the pressure of 10 or 12 atmospheres. The gas gives the H_2O a pleasant, pungent, slightly acid taste, and by its escape, when exposed to the air, produces a brisk effervescence.* In beer, ginger-pop, cider, wine, etc., the CO_2 is produced by fermentation.† The gas escapes rapidly through cider and wine, and so produces only a sparkling; while in a thick, viscid liquid, like beer, the bubbles are partly confined, and hence cause it to foam and froth. In canned fruits, catsup, etc., the "souring" of the vegeta-

since it will not burn, and it is said that miners will even venture "where the air is so foul that the candles go out, and are then re-lighted from the coal on the wick by swinging them quickly through the air, when they burn a little while and then go out, and are re-lighted in the same way."

* Pass a current of CO_2 through a gill of water. Add a few drops of blue litmus-solution. It will immediately redden. Boil the water, when the gas will escape and the water become blue.

† Dissolve an oz. of sugar in 10 times its weight of water. Put it in a flask like that shown in Fig. 24, and add a little fresh brewer's yeast. If kept warm, in a short time it will give off CO_2, which may be tested.

bles produces CO_2, which sometimes drives out the cork or bursts the bottles with a loud report.

Liquid CO_2.—By a pressure of 36 atmospheres, at a temperature of 32°, CO_2 becomes a colorless liquid, very much like H_2O. When this is exposed to the air it evaporates so rapidly that a portion is frozen into a snowy solid which burns the flesh like red-hot iron. By means of solid CO_2, Hg can be readily frozen. When mixed with ether, and evaporated under the exhausted receiver of an air-pump, a cold of —148° may be produced. (See *Physics*, p. 191.)

Ventilation.—The relation of CO_2 to life is most important, and cannot be too often dwelt upon. We exhale constantly this dangerous gas, and if fresh air is not furnished continuously we are forced to rebreathe that which our lungs have just expelled.* The languor and sleepiness we feel in a crowded assembly, are the natural effects of the vitiated atmosphere.† The idea of drinking in at every breath the exhalations that load the air of a crowded, promiscuous assembly-room, is a most disgusting one. We shun impurity in every form; we dislike to wear the clothes of another, or to eat from the same dish; we shrink from contact with the filthy, and yet sitting in the same room inhale their polluted breath. Health and cleanliness alike require that we should care-

* It is a fact, as poetical as it is characteristic, that when the air comes forth from the lungs it is charged with the seeds of disease; yet, as it passes out, it produces all the tones of the human voice, all songs, and prayers, and social converse. Thus the gross and deadly is by a divine simplicity made refined and spiritual, and caused to minister to our highest happiness and welfare.

† It should be noted that the deleterious effects of ill-ventilation arise not only from the presence of CO_2, but from the organic particles given off in the breath and exhaled from the skin. (See *Physiology*, page 93.) Rebreathed air is a fruitful source of consumption and scrofula.

fully ventilate public buildings, school-rooms, and sleeping apartments.*

Fig. 28.

Testing the currents of air to and from flame.

Carbonic Oxide, CO, is a colorless, almost odorless gas, and has never been liquefied. It burns with a pale blue flame, absorbing an atom of O from the air, and be-

* Two openings are necessary to ventilate a room. To illustrate this, set a lighted candle in a plate of water, as shown in Fig. 28. Cover it with an open jar, over the neck of which is placed a common lamp-chimney. The light will soon be extinguished on account of the consumption of O, and the formation of CO_2. Raise the jar at one side a trifle above the water, and the candle, if re-lighted, will burn steadily—fresh air coming in below, and the refuse passing off at the top. Replace the jar, and as the candle is flickering, insert in the chimney a slip of card, thus dividing the passage, when the light will brighten again. Hold a bit of smouldering touch-paper (page 45) at the top, and the smoke will show two opposite currents of air established in the chimney. Mines have been ventilated in this way by dividing the shaft. More commonly, however, they have two shafts at a little distance apart.

coming CO_2. It is seen burning thus in our coal-stoves, and at the tops of tall furnace-chimneys. It is often formed abundantly on account of the action of heated carbon on CO_2. When air enters at the bottom of a clear fire, CO_2 is formed at once; but this gas passing through the hot embers takes up a further quantity of C, becoming changed into CO:* $C+CO_2=2CO$, the volume of the gas being exactly doubled in bulk thereby. CO is a deadly poison, and escaping from coal-fires in a close room has often produced death. Both CO and CO_2 leak through the pores of cast Fe when heated, and still further injure the air of our houses and necessitate ventilation. The offensive odor which comes out on opening the door of our coal-stoves is caused by the compounds of S mixed with the CO.

Marsh Gas.—*Light Carburetted Hydrogen*, CH_4 (see p. 200).—This we have already spoken of under CO_2, as the dreaded fire-damp of miners. It is colorless, tasteless, odorless, and burns with a yellowish flame. It is formed in swamps and low marshy places by the decomposition of vegetable matter, and on stirring the mud beneath will be seen bubbling up through the water. It may be collected in the manner shown

Fig. 29.

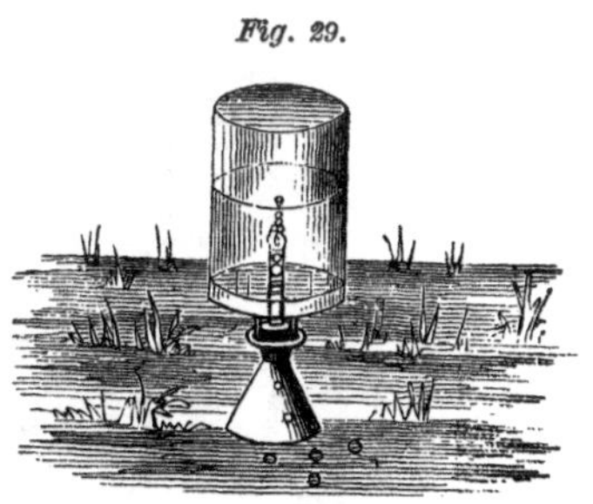

Collecting Marsh-gas.

* This fact is of great importance, since thereby much heat is wasted. Stoves are sometimes so constructed as to admit fresh air just above the grate, thus consuming this gas.

Fig. 30.

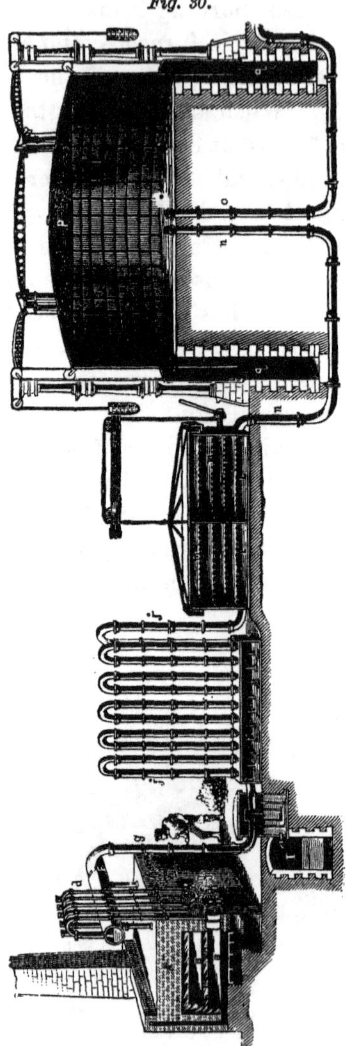

Manufacture of Coal-gas.

in Fig. 29. It rises from the earth in great quantities at many places. At Fredonia, N. Y., it is used in lighting the village. At Kanawha, Va., it was until lately employed as fuel for evaporating the brine in the manufacture of salt. In the oil-wells of Pennsylvania, it frequently bursts forth with explosive violence, throwing the oil high into the air.

Olefiant Gas.—*Heavy Carburetted Hydrogen*, C_2H_4.—This is a colorless gas, with a sweet, pleasant odor, and burns with a clear white light.* It may be easily prepared by heating in a large retort a mixture of one part of alcohol with two of H_2SO_4.

Coal Gas is a very variable mixture. The

* Fill a tall jar one-third full of olefiant gas, and the remainder with chlorine gas. On lighting, the mixture will burn with a dense cloud of smoke. HCl is the product of the combustion.

proportion of olefiant gas and hydrocarbons having a similar composition gives whiteness to the flame; while the H and CH_4 have little illuminating power, and serve mainly as carriers of the more valuable gases (MILLER). Bituminous coal is heated in large iron retorts, *B*, until only coke is left and the volatile constituents are driven off. Among them are coal-tar, H_3N, CO_2, CO, N, compounds of S, CH_4, and C_2H_4.* This mixture is led through the curved pipes, *d*, beneath the H_2O in the *hydraulic main*, F; along the tube, *g*, to the *tar cistern;* thence up and down the *condenser*, *j*. On the way it becomes cooled and loses its coal-tar, ammoniacal salts,† and liquid hydrocarbons. Lastly it is passed over lime, *Lm*, which absorbs the CO_2 and the H_2S.‡ The remaining gases form the mixture we call "gas." This is collected in the gasometer, P, the weight of which forces it through all the little gas-pipes, and up to every jet in the city.

Coal-gas is very poisonous, and even in small quantities exceedingly deleterious. When mixed with air it explodes with great violence. Its unpleasant odor, though often annoying, is a great protection, as we are thereby warned of its presence.

Cyanogen,§ Cy=CN.—**Preparation.**—As N and C

* None of these substances exists in coal. They are formed by the action of heat, which causes the H, C, O, N and S to combine and make a multiplicity of compounds.

† The H_3N is neutralized by HCl, thus forming chloride of ammonium (sal-ammoniac, H_4N,Cl). On evaporation, the tough, fibrous crystals of the salt are obtained. (See page 135.)

‡ The removal of the sulphur compounds is especially important, since, when burned, they furnish sulphurous and sulphuric acids, which are very injurious to books, paintings, and furniture.

§ The term cyanogen means "blue producer;" this gas being the characteristic constituent of Prussian blue.

do not combine directly, this gas is obtained in an indirect way. Mix the parings of horns, hides, etc., with pearlash (potassium carbonate) and iron filings, and heat in a close vessel. The N and C of the animal substances, in their nascent state, will combine, forming Cy; this uniting with the Fe and K will produce the beautiful yellow crystals of potassium ferro-cyanide (yellow prussiate of potash). From this salt the mercury cyanide is made, which when heated decomposes into Hg and Cy.

Properties.—Cy is a transparent, colorless gas, with a penetrating odor. It burns with a characteristic rose-edged purple flame, and is exceedingly poisonous. It is very interesting from the fact that, though a compound, it unites directly with the metals like the elements I, B, etc. It is therefore called a *compound radical* (root). We shall find this subject of great importance in Organic Chemistry.

Hydrocyanic Acid, HCy.—Prussic acid, as it is commonly called, is a fearful poison. A single drop on the tongue of a large dog is said to produce instant death. H_3N, cautiously inhaled, is its antidote. Its bitter flavor is detected in peach blossoms, the kernels of plums or peaches, bitter almonds, and the leaves of wild cherry.

Fulminic Acid (*fulmen*, a thunderbolt).—This compound of Cy is known only as combined with the various metals forming fulminates, which are remarkably explosive. Fulminating mercury was used to fill the bombs with which the life of Napoleon III. was attempted in 1858. It is employed in making gun-caps. A drop of gum is first put in the bottom of the cap, over which is sprinkled a little fulminating mercury, and this is sometimes covered with varnish to protect it from any moisture.

COMBUSTION.

COMBUSTION, in general, is the rapid union of a substance with O, and is accompanied by heat and light.*

The Igniting Point of any substance is the temperature at which "it catches fire." We elevate the heat of a small portion to the point of rapid union with O, and that part in burning will give off heat enough to support the combustion of the rest.—*Example:* In making a fire, we take paper or shavings, which being poor conductors of heat, and exposing a large surface to the action of O, are easily raised to the required temperature. Having thus obtained sufficient heat to start the combustion of chips or pine sticks, we gradually increase it until there is enough to ignite the coal or wood.

Chemistry of a Fire.—Our fuel and lights, such as wood, coal, oil, tallow, etc., consist mainly of C and H, and are, therefore, called *hydrocarbons.* In burning they unite with the O of the air, forming H_2O and CO_2. These both pass off, the one as a vapor, the other as a gas. In a long stove-pipe, the H_2O is sometimes condensed, and drips down, bringing soot upon our carpets. Ashes comprise the mineral matter contained in the fuel, united with some of the CO_2 produced in the fire. When we first put fuel in the stove, the H is liberated with some C, in the form of marsh or olefiant gas.

* There are forms of combustion known to the chemist which are not oxidation; as the union of S and Cu. (See page 97, note.)

This burns with a flame. Then, the volatile H having passed off, we have left the C, which burns as coal merely. In maple there is much more C than in pine, so it forms a good "bed of coals." In the burning of fuel there is no annihilation; but the H_2O, CO_2, and the ashes, weigh as much as the wood and the O that combined with it. No matter how rapidly the fire burns, even in the blaze of the fiercest conflagration, the elements unite in exact atomic weights.

C is most wisely fitted for fuel, since the product of its combustion is a gas. Were it a solid, our fires would be choked, and before each supply of fresh fuel we should be compelled to remove the ashes. In the case of a candle or lamp it would be still more annoying, as the solid product would fall around our rooms. Still another useful property is the infusibility of C. Did C melt like Z or Pb on the application of heat, how quickly in a hot fire would the coal and wood run down through the grate and out upon the floor in a liquid mass!

Fig. 31.

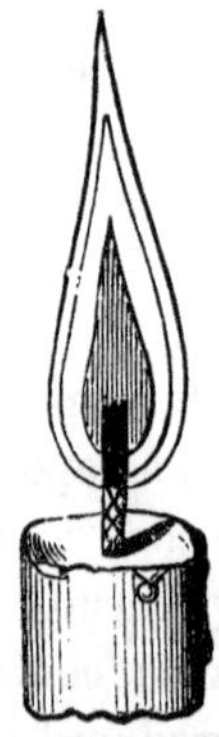

Form of flame.

Chemistry of a Candle. — Flame is burning gas. A candle is a small "gas-work," and its flame is the same as that of a "gas-burner." First, we have a little cupful of tallow melted by the heat of the fire above. The ascending currents of cool air which supply the light with O also keep the sides of the cup hard, unless the wind blows the flame downward, when the banks break, there is a *crevasse*, and our "candle runs down." Next, the melted tallow is carried by capillary attraction up the small tubes of the wick into the flame. There it is turned into gas by the heat. Flame is always hollow, and at the center, near the wick, is the gas just formed. If a match be placed across a light, it will burn off at each side, in the ring of the flame, while the center will be unblackened.* The gas may be conducted out of the flame by a small pipe, and burned at a little distance from the candle. Flame is hollow because there is no O at the center. The gas floats outward from the wick. It comes in contact with the O of the air, and the H, requiring least heat to unite, burns first, forming H_2O. This produces heat enough to make the tiny particles of C, floating around in the flame of burning H, white-hot.† They each send out a delicate wave of light, and passing on to the outer part, where there is more O, burn, forming CO_2. The flame is blue at the bottom, because there is so much O at that point that the H and O burn together, and so give little light. The H_2O may be condensed on any cold surface. The CO_2 may be tested by passing the invisible vapor of a candle through lime-water. The wick of a candle does not burn because of the lack of O at the center. It, how-

Fig. 32.

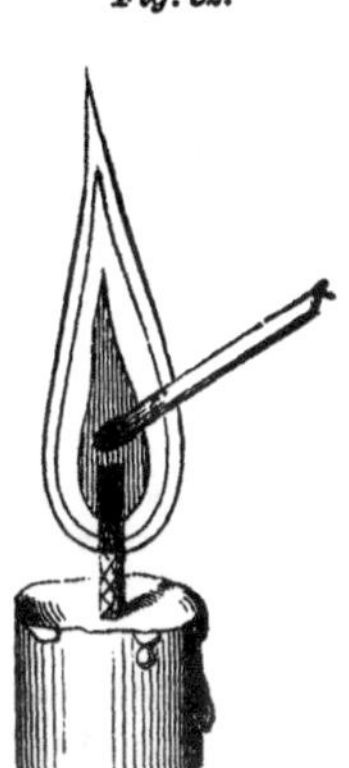

Match in flame; the S and P being unconsumed.

* Take a sheet of white paper and thrust it quickly down upon the flame of a candle or lamp. It will burn in a ring, and when the paper is removed the center will be found unblackened.

† Frankland has shown that the intensity of a flame, in general, is determined by the density of the gas: thus a jet of H burning under a pressure of ten atmospheres will furnish sufficient light to read a newspaper at a distance of two feet.

ever, is charred, as all the volatile gas is driven off by the heat. If a portion falls over to the outer part, where

Fig. 33.

Testing the CO_2 of a flame by drawing the gas through lime-water.

there is O, it burns as a coal. If we blow out a candle quickly, the gas still passes off, and we can relight it with an ignited match held at some distance from the wick. The tapering form of the flame is due to the currents of air that sweep up from all sides toward it. The candle must be snuffed, because the long wick would cool the blaze below the igniting point of C and O, and the C would pass off unconsumed. A draught of air, or any cold substance thrust into the flame, produces the same result, and deposits the C as soot. Plaited wicks are sometimes used, which, being thin, fall over to the outside and burn, requiring no snuffing.

Chemistry of a Lamp.—A chimney confines the hot air, and makes a draught of heated O to feed the flame. A flat wick is used, as it presents more surface to the action of the O. Argand lamps have a hollow wick which admits O into the center of the blaze. The film which gathers on a chimney when we first light a lamp, is the H_2O produced in the flame, condensed on the cold glass. A pint of oil forms a full pint of H_2O. Spirits of turpentine, tar, pine-wood, etc., contain an excess of C, and not enough H to heat it to the igniting point. These, therefore, produce clouds of soot. Alcohol contains an excess of H and little C, hence it gives off great heat and little light.

Fig. 34.

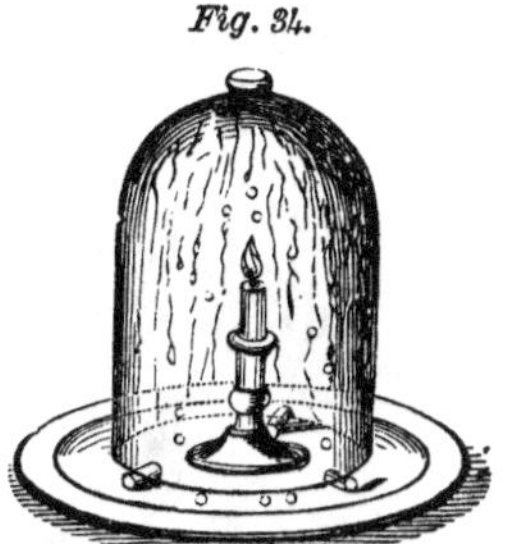

H_2O *condensed from a flame.*

Fig. 35.

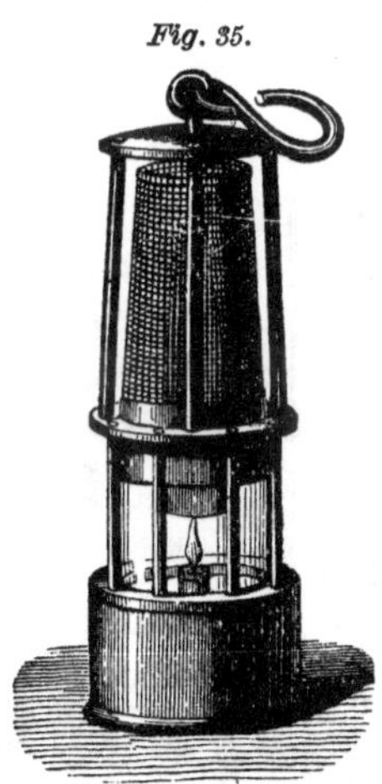

Davy's Safety Lamp.

Davy's Safety Lamp, used by miners, consists of an ordinary oil-lamp, surrounded by a cylinder of fine wire-gauze. When it is carried into an atmosphere containing the dreaded fire-damp, the flame enlarges and becomes pale, and when the quantity increases, the gas will quietly burn on the inside of the cylinder.* There is no danger of an explosion so long as the gauze

* The principle of the lamp can be illustrated by holding a fine wire-gauze over the flame of a candle or lamp (Fig. 36). The flame will not pass through, since the wire will conduct away the heat and so reduce the temperature below the igniting point. A jet of gas, issuing at a low temperature, may be lighted on either side of the gauze at pleasure.

remains perfect.* Through carelessness, however, fearful accidents have occurred. Miners become extremely negligent, and an account is given of an explosion, in which about a hundred persons were killed, caused by a lamp being hung on a nail by a hole broken through the wire-gauze.

Fig. 36.

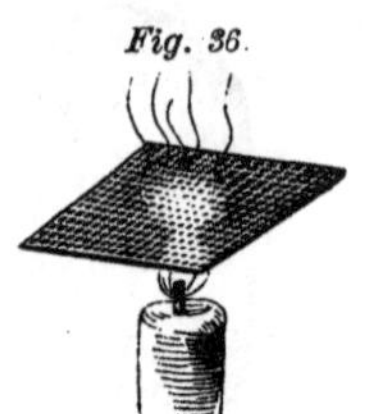

Wire gauze over flame.

Bunsen's Burner is used in the laboratory. It consists of a jet of gas, *a*, surrounded by a metal tube, *c*, at the bottom of which are openings, *b*, for the admission of air. The gas passes up the tube, mingles with the air, and burns at the top without smoke. The O is supplied in sufficient quantity to burn the H and C simultaneously; hence there is great heat with little light and no smoke.

Fig. 37

Bunsen's burner.

The Oxy-hydrogen Blow-pipe is so constructed that a jet of O is introduced into the center of one of burning H, thus producing a *solid* flame. A watch-spring will burn in it with a shower of

* At such a time, however, the wise miner will leave the place of danger, lest the metal should melt and the fire escape to the gas, when an explosion would ensue.

sparks. Pt, the most infusible of metals, will readily melt. In the common flame, as we have seen, the little

Fig. 38.

The Oxy-hydrogen Blow-pipe.

particles of solid C, heated by the burning H, produce the light. As there is no solid body in the blow-pipe flame, it is scarcely luminous. If, however, we insert in it a bit of CaO, or MgO, a dazzling light is produced. This is called the "Drummond." "Lime," or "Calcium"

Light, and with a properly arranged reflector has been seen at a distance of one hundred and eight miles in broad sunlight.

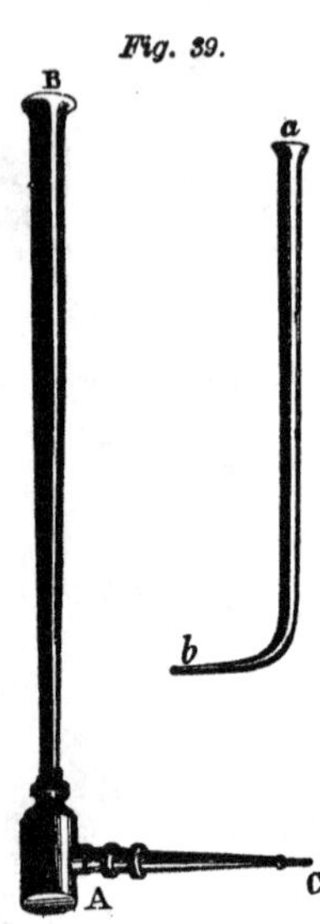

Common Blow-pipe.

Mouth Blow-pipe.—In the common blow-pipe, used by jewelers and mineralogists, a current of air from the lungs is thrown across the light just above the wick. The flame loses its brilliancy and is driven one side in the form of a cone (Fig. 40). Three parts are clearly visible. In the center is a blue cone ending at *a*; outside of this is a whiter and more luminous one terminating at *b*; and beyond this a pale yellow flame, *c*. The blue cone is caused by the excess of O from the blow-pipe burning the C and H at the same time. The luminous cone contains C in excess, which being heated gives out light. The combustible vapor at this point, hot and ready to combine, will take O from any substance exposed to it, and is therefore called the *reducing flame*. The outer en-

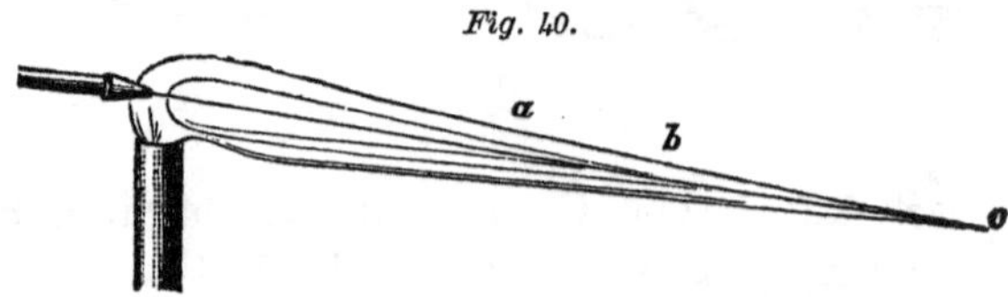

velope contains the O thrown from the lungs, borne forward by the jet of flame, and highly heated by it. It will unite with a metallic body, and is therefore called the *oxidizing flame*.—*Example:* Hold a copper cent in the blaze of an alcohol lamp; in the "reducing flame" its rust, copper oxide, will be cleaned off, and the metal will shine as brightly as if just from the mint. In the "oxidizing flame" a film of copper oxide will be formed over the surface, and as we move the cent the most beautiful play of colors will flash from side to side.*

Extinguishing Fires.—Blowing on a candle or lamp extinguishes it, because it lowers the heat of the flame below the igniting point of the gases.† Fires are put out by H_2O partly for the same reason, and also because it envelops the wood and shuts off the air. If a person's clothes take fire, the best course is to wrap him in a blanket, carpet, coat, or even in his own garments. This smothers the fire by shutting out the O. Great care should be taken in a fire not to open the doors or windows, so as to cause a draught of air. The entire building may burst into a blaze, when the fire might have languished for want of O, and so have been easily extinguished.

Spontaneous Combustion.—Sometimes chemical changes take place in combustible substances, whereby heat enough is generated to cause ignition. CaO occasionally absorbs H_2O, so as to set fire to wood in contact with it. Fresh-burned charcoal has the power of absorbing gases in its pores so rapidly as to become ignited.

* Introduce a small piece of common flint-glass tube into the *reducing flame*. The glass will become opaque and black, because the Pb will be reduced from the transparent form of oxide to the opaque condition of metal. When this has happened, place the black portion just in front of the oxidizing flame. The discoloration will slowly disappear, and the Pb will recombine with O from the air and the glass again become transparent.

† Sometimes, also, the flame is driven off by the mere force of the breath.

Heaps of coal take fire from the iron pyrites contained in them, which is decomposed by the moisture of the air. The waste cotton used in mills for wiping oil from the machinery, when thrown into large heaps, often absorbs O from the air so rapidly that it bursts into a blaze.

PRACTICAL QUESTIONS.

1. Why does not blowing *cold* air on a fire with a bellows extinguish it?
2. Why will pine-wood ignite more easily than maple?
3. Why is fire-damp more dangerous than choke-damp?
4. Represent the reaction in making CO_2, showing the atomic weights, as in the preparation of O on page 28.
5. Should one take a light into a room where the gas is escaping?
6. What causes the difference between a No. 1 and a No. 4 pencil?
7. Why does it dull a knife to sharpen a pencil?
8. Why is slate found between seams of coal?
9. Why was the coal hidden in the earth?
10. Where was the C, now contained in the coal, before the Carboniferous age?
11. Must the air have then contained more plant food? (p. 98.)
12. What is the principle of the aquarium?
13. What test should be employed before going down into an old well or cellar?
14. What causes the sparkle of wine, and the foam of beer?
15. What causes the cork to fly out of a catsup bottle?
16. What philosophical principle does the solidification of CO_2 illustrate?
17. Why does the division in the chimney shown in Fig. 28 produce opposite currents?
18. What causes the unpleasant odor of coal-gas? Is it useful?
19. What causes the sparkling often seen in a gas-light?
20. Why does H in burning give out more heat than C?
21. Why does blowing on a fire kindle it, and on a lighted lamp extinguish it?
22. Why can we not ignite hard coal with a match?

23. What causes the dripping of a stove-pipe?
24. Why will an excess of coal put out a fire?
25. Why do not stones burn as well as wood?
26. Why does not hemlock make "a good bed of coals?"
27. What adaptation of chemical affinities is shown in a light?
28. Is there a gain or a loss of weight by combustion?
29. Why does snuffing a candle brighten the flame?
30. Why is the flame of a candle red or yellow, and that of a kerosene oil-lamp white?
31. Why is it beneficial to stir a wood-fire, but not one of anthracite coal?
32. Why will water put out a fire?
33. What should we do if a person's clothes take fire?
34. Ought the doors of a burning house to be thrown open?
35. Why does a street gas-light burn blue on a windy night? Is the light then as intense? The heat?
36. Why does not the lime burn in a calcium-light?
37. Why is a candle-flame tapering?
38. Why does a draught of air cause a light to smoke?
39. What makes the coal at the end of a candle-wick?
40. Which is the hottest part of a flame?
41. Why does not a candle-wick burn?
42. How does a chimney enable us to burn without smoke highly carboniferous substances like oil?
43. How much CO_2 in 200 lbs. of chalk?
44. What weight of CO_2 in a ton of marble?
45. What is the difference between marble and chalk? (See page 139.)
46. Why does not a cold saucer held over an alcohol flame blacken, as it does over a candle or gas-light?
47. Could a light be frozen out, *i. e.*, extinguished, by merely lowering the temperature?
48. How much CO_2 is formed in the combustion of one ton of C?
49. What weight of C is there in a ton of CO_2?
50. How much O is consumed in burning a ton of C?
51. What weight of sodium carbonate (Na_2CO_3, "carbonate of soda") would be required to evolve 12 lbs. of CO_2?
52. How much CO_2 will be formed in the combustion of 30 grs. of CO?
53. What weight of hydrogen sodium carbonate ($HNaCO_3$, "bicarbonate of soda") would be required to evolve 12 lbs. of CO_2?

54. Write in double columns the different properties of carbonic anhydride and carbonic oxide; thus,

CO_2 is	CO is
1, non-inflammable.	1, inflammable.

THE ATMOSPHERE.

The "air we breathe" consists of N, O, CO_2, and watery vapor. The first composes $\frac{4}{5}$, the second $\frac{1}{5}$, the third about $\frac{4}{10,000}$, and the last a variable amount.* A very clear idea of the proportion of these several constituents may be formed by conceiving the air, not as now dense near the surface of the earth, and gradually becoming rarefied as we ascend to its extreme limit of perhaps 500 miles, but of a density throughout equal to that which it now possesses near the earth. The atmosphere would then be about five miles high. The vapor would form a sheet of H_2O over the ground five inches deep, next to this the CO_2 a layer of 13 feet, then the O a layer of one mile, and last of all the N one of four miles.—GRAHAM. In this arrangement we have supposed the gases to be placed in the order of their specific gravity. The atmosphere is not thus composed in fact, the various gases being equally mingled throughout, in accordance with a principle called the "*Law of the diffusion of Gases.*" If we throw a piece of lead into a brook, it will settle instantly to the bottom by the law of gravitation, and will remain there by the law of inertia. But if we throw into the atmosphere a quantity of CO_2, it will sink for an instant, then immediately begin to mingle

* The N and O form so large a part, that they are considered in ordinary calculation to compose the whole atmosphere.

with the surrounding air, and soon become dissipated.—*Example:* If we invert an open-mouthed bottle full of H over another full of CO_2, the H, light as it is, will sink down into the lower jar; and the CO_2, heavy as it is, will rise into the upper jar; and in a few hours the gases will be found equally mixed. By this law the proportion of the elements of the atmosphere is the same everywhere, and has not varied within historic times. Samples have been analyzed from every conceivable place, from polar and torrid regions, from prairies and mountain-tops, from balloons and mines, from crowded capitals and lonesome forests, and even from bottles found sealed up in the ruins of Herculaneum, and the result is almost exactly the same. These gases do not form a chemical compound, but a mere mechanical mixture,* and they are as distinct in the air as so many grains of wheat and corn mingled in a measure.

Fig. 41.

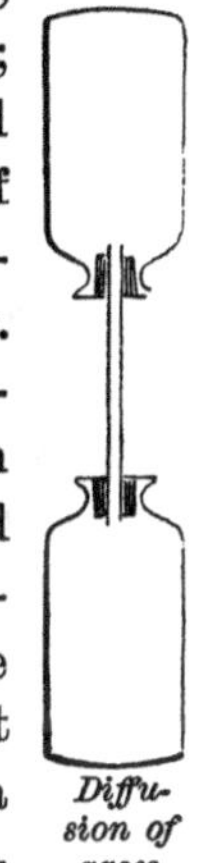

Diffusion of gases.

Each of the constituents of the air has its separate use and mission. The action of O and N we have already seen.

Uses of CO_2.—Carbonic acid bears the same relation to vegetable that O does to animal life. The leaf—the

* "To illustrate the difference between a mechanical mixture and a chemical compound, mix powdered S and filings of Cu. The color of the S as well as that of the Cu will disappear, and to the unaided eye will present a uniform greenish tint; with the microscope, however, the particles of Cu may be seen lying by the side of those of S; and we can wash away the lighter S with H_2O, leaving the heavier Cu behind. Here no *chemical action* has occurred; the S and Cu were only *mechanically mixed*. If we next gently heat some of the mixture it soon begins to glow, and on examining the mass we find that both the Cu and the S have disappeared as such, that they cannot be distinguished even with the most powerful microscope, and that in their place we have formed a black substance possessing properties entirely different from those possessed either by the Cu or the S."—ROSCOE.

plant-lungs—through its million of little *stomata*, or mouths, drinks in the CO_2. In that minute leaf-laboratory, by the action of the sunbeam, the CO_2 is decomposed,* the C being applied to build up the plant, and the O returned to the air for our use. Plants breathe out O as we breathe out CO_2. We furnish vegetables with air for their use, and they in turn supply us. There is thus a mutual dependence between the animal and the vegetable world. Each relies upon the other. Deprived of plants we should soon exhaust the O from the air, supply its place with CO_2, and die; while they, removed from us, would soon exhaust the CO_2, and die as certainly. We pollute the air while they purify it. Each tiny leaf and spire of grass is thus imbibing our foul breath, and returning it to us pure and fresh.† This in-

* "In order to decompose carbonic acid in our laboratories, we are obliged to resort to the most powerful chemical agents, and to conduct the process in vessels composed of the most resisting materials, under all the violent manifestations of light and heat, and we then succeed in liberating the carbon only by shutting up the oxygen in a still stronger prison; but under the quiet influences of the sunbeam, and in that most delicate of all structures, a vegetable cell, the chains which unite together the two elements fall off, and, while the solid carbon is retained to build up the organic structure, the oxygen is allowed to return to its home in the atmosphere. There is not in the whole range of chemistry a process more wonderful than this. We return to it again and again, with ever increasing wonder and admiration, amazed at the apparent inefficiency of the means, and the stupendous magnitude of the result. When standing before a grand conflagration, witnessing the display of mighty energies there in action, and seeing the elements rushing into combination with a force which no human agency can withstand, does it seem as if any power could undo that work of destruction, and rebuild those beams and rafters which are disappearing in the flames? Yet in a few years they will be rebuilt. This mighty force will be overcome; not, however, as we might expect, amidst the convulsion of nature, or the clashing of the elements, but silently, in a delicate leaf waving in the sunshine."—COOKE.

† From this statement it is evident that the foliage of house-plants must be healthful. Moreover, there is some reason to believe that the O which they exhale is highly ozonized, and therefore of great value in destroying miasmic germs. We should remember, however, that flowers exhale CO_2; and the odor of certain plants, and the pollen of others, are very injurious. Plants and flowers, which to one person are innocuous, are to another detrimental. Thus

Fig. 43.

Apparatus arranged to catch the O evolved from a sprig of leaves.

terchange of office is so exactly balanced, that, as we have seen, the proportion of CO_2 and of O, in the open air, never varies.*

the fragrance of new-mown grass, which is so agreeable to some, produces in others what is termed the *hay-fever;* due, it is said, to the pollen of the grass. Each family, therefore, must determine for itself what should be excluded from its collection. It is evident that flowerless plants, like the ivy, etc., are harmless, while the cheerfulness given to an apartment by even a few pots of flowers on a window-bench, should induce one to take some trouble in order to make a selection which will not only beautify but purify the room.

* "Two hundred million tons of coal are now annually burned, producing six hundred million tons of CO_2. A century ago, hardly a fraction of that amount was burned, yet this enormous aggregate has not changed the proportion in the least."—YOUMANS.

Plants Store up Solar Force.—The sunbeam, which is thus strong enough to wrench apart the C and O, sends out the O full of potential force, and, by its energy, builds up the plant. The force of the sunbeam is then latent in the vegetable structure. The sun shining on a meadow causes the grass to grow. If the hay made from it be eaten by an animal, the same amount of force will be liberated as was received from the sun. A tree towers upward through a century of sunshine. When burned, it sets free as much force as was needed to perfect its growth. A bushel of corn, then, represents not alone so much C, H, and O, but also an amount of sun-force which is available for any purpose to which we wish to apply it. (See *Conclusion.*)

Animals Spend Solar Force.—In the process of digestion the force stored in the plant is transferred to the animal, is given out by its muscles on their oxidation and produces motion, heat, etc. H_3N, CO_2, and H_2O are decomposed by the plant and organized into complex molecules (see p. 182), full of potential force. The animal oxidizes the organic molecules, and breaks them up into H_3N, CO_2, and H_2O again—simple molecules robbed of force which the animal has used. Thus the plant builds up and the animal tears down. The plant garners in the sunbeam and the animal scatters it again. The plant reduces and the animal oxidizes.

Uses of Watery Vapor.—We have already seen the uses of H_2O. As vapor, it is everywhere present and ready to supply the wants of animals and plants. Were the air perfectly dry, our flesh would become shriveled like a mummy's, and leaves would wither as in an African simoom. Rivers and streams flow to the ocean; yet all their fountains are fed by the currents that move in the air above us. H_2O rises as vapor, flows on to colder regions, falls as rain, dew, snow, or hail, and then working as it goes whatever it finds to do, moistening a plant or turning a water-wheel, wends its way back to the ocean. Thus Niagara itself must first rise to the clouds as vapor before it can fall as a cataract.

Permanence of the Atmosphere.—The elements of the air unite to form HNO_3 only by the passage of electricity, and then in minute quantities. If they combined more readily we should be constantly exposed to a shower of this corrosive acid that would be destructive to all vegetation, clothing, and even our bodies themselves.—O,* N, and CO_2 are reduced from their gaseous form only by an apparatus specially made for the purpose, and under circumstances which could rarely, if ever, occur in Nature. These substances are therefore constantly in the condition promptly to supply the demands of animals and plants.—Watery vapor, on the contrary, is deposited as dew or rain by the slightest change of temperature; this readiness of condensation is equally necessary to meet the wants of animal and vegetable life.—The permanence of the air produces all the uniformity of sound. Were the proportions of the atmosphere to change, all "familiar voices" would become strange and uncouth, while the harmonies of music would shock us with unwonted discord.† Each element of the air is adapted to a special work, and all are fitted to the present order of nature.

* The liquefaction of the so-called "permanent gases," N, O, H, etc., is the great chemical triumph of 1877. It was accomplished by Cailletet of Paris and Pictet of Geneva, almost simultaneously.

† If, by some means, the air of a concert-room could be changed to H, for instance, the bass voices would become irresistibly comic and shrill, while the tenor would emulate railway whistles.

THE HALOGENS.

Chlorine..	Symbol, Cl;	Atomic Weight, 35.5;	Specific Gravity,	2.43
Iodine....	" I;	" " 127.;	"	4.94
Bromine..	" Br;	" " 80.;	" (at 32°),	3.18
Fluorine..	" F;	" " 19.;	"	1.31

These four elements are closely allied, and are known as the *halogens*, from *hals*, salt, because they form a class of compounds (Haloids) which resemble common salt (NaCl).*

CHLORINE is named from its green color. It is chiefly found in salt, of which it forms 60 per cent. It is pre-

Fig. 43.

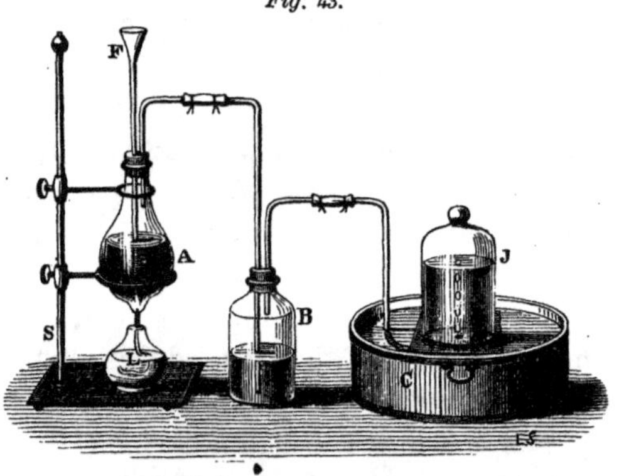

Preparing Cl.

* In comparing the halogens with one another, the chemical activity of F, which has the smallest atomic weight, is the most powerful; next in the order of activity is Cl, then Br, and, lastly, I, the atomic weight increasing as the chemical energy declines. Cl is gaseous, Br, liquid, and I solid. The specific gravity, the fusing point, and the boiling point, rise as the atomic weight increases. The halogens combine energetically with the metals, and, when united with the same metal, furnish compounds which are *isomorphous;* that is to say, they all crystallize in the same form—potassium fluoride, chloride, bromide, and iodide, for example, all crystallize in cubes. Each, also, forms with H a soluble, powerful acid—HCl, HI, HBr, HF.

pared by heating NaCl with MnO_2, H_2SO_4, and H_2O.* This mixture liberates the gas in great quantities. Cl is heavier than common air, and hence may be collected by displacement, as in the preparation of CO_2, or a solution of the gas may be obtained by using the apparatus shown in Fig. 43, while the excess is gathered in a receiver.

Properties.—Cl has a greenish-yellow color, and a peculiarly disagreeable odor. It produces a suffocating cough, which can be relieved by breathing ammonia or ether. Arsenic, Dutch gold-leaf, phosphorus, etc., combine with it so rapidly as to inflame. Powdered antimony slowly dropped into it produces a shower of brilliant sparks. Cold water absorbs about twice its volume of the gas, which, in the sunlight, turns to hydrochloric acid (HCl). Cl has such a powerful affinity for H, that it will even attract it out of a moist organic body, and form HCl. It acts thus upon turpentine, depositing its C in great flakes of soot. It discharges the color of indigo, ink, wine, etc., almost instantaneously. It has no effect on printers' ink, the coloring matter of which contains no H. (See p. 222.)

Fig. 44.

Turpentine in Cl.

Uses.—*Bleaching.*—In domestic bleaching the cloth is first boiled with strong soap, to dissolve the grease and wax, and then laid upon the grass, being frequently wet to hasten the action of the air and sun. The dew seems to have a peculiar influence, while the corrosive ozone of the atmosphere doubtless aids in the process. The H of the coloring matter unites with the O of

* The chemical reaction is as follows:

Manganese Dioxide		Sodium Chloride		Sulphuric Acid		Manganese Sulphate		Hydrogen Sodium Sulphate		Water		Chlorine
MnO_2	+	$2NaCl$	+	$3H_2SO_4$	=	$MnSO_4$	+	$2HNaSO_4$	+	$2H_2O$	+	Cl_2.

the air or dew, forming H_2O, and destroying the coloring compound.*

The method of bleaching on a large scale is as follows: The cloth is well washed, and boiled in water with strong alkalies, to remove the grease, etc.; next it is passed through a solution of chloride of lime, and lastly through diluted H_2SO_4. In this step the acid unites with the lime, and sets free the Cl, which in turn combines with the H of the coloring matter, forming HCl, and thus bleaches the cloth. "About twenty-four hours are required for this process, and the cost is not quite a cent per yard." Paper-rags are bleached in the same way in paper-mills.†

Disinfectant.—Cl is a powerful disinfectant. It breaks up the offensive substance by uniting with its H, as in bleaching. Other disinfectants, as burnt paper, sugar, etc., only disguise the ill odor by substituting a stronger one. In the sick-room Cl is set free from chloride of lime (bleaching powder) by exposing it to the air in a saucer with a little H_2O. The gas soon passes off, though the process may be hastened by adding a few drops of dilute acid. Chloride of lime is, therefore, of

* This was essentially the process long pursued in Holland, where linens were formerly carried for bleaching; hence the term "Holland linen," still in use. The H_2O about Haarlem was thought to have peculiar properties, and no other could compete with it. Cloths sent there were kept the entire summer, and were returned in the fall. Later a similar plan was adopted in England. But the vast extent of grass-land required, the time occupied, and the temptation to theft, made the process extremely tedious and expensive. The statute laws of that time abound in penalties for cloth stealing. It is estimated that all the men, women, and children in the world could not, by the old way, bleach all the cloth that is now used.

† Stains can be removed from *uncolored cloth* by "Labarraque's Solution," a compound of Cl, which can be obtained of any druggist. Place the cloth in this liquid, and if the stain is obstinate, pour on a little boiling H_2O, or place it in the sun for some hours. Then rinse thoroughly in cold H_2O, and dry.

great service for disinfecting all places exposed to any noxious or unpleasant effluvia. Hospitals and rooms in which persons have died of a contagious disease are purified by placing in them pans full of a mixture which is disengaging Cl in large quantities.

Compounds.—*Hydrochloric Acid, Muriatic Acid,* HCl.—When Cl and H are mixed in the dark and exposed to the direct sunlight they unite with an explosion. In the arts HCl is prepared from H_2SO_4 and NaCl. The reaction is as follows: $NaCl + H_2SO_4 = HCl + HNaSO_4$.

Fig. 45.

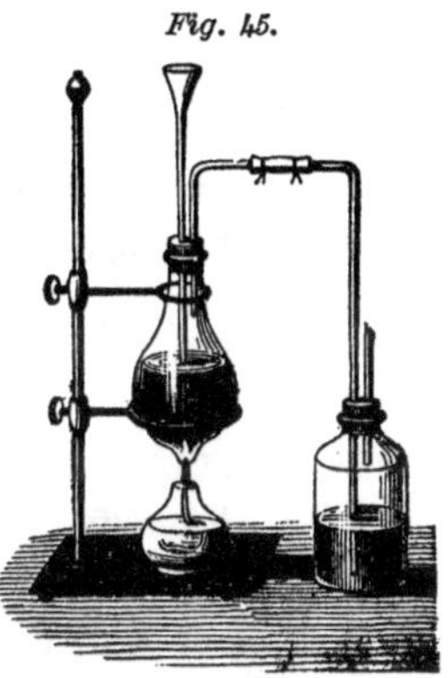

Preparing HCl.

Properties.—It is an irrespirable, irritating, acid gas, with an intense attraction for H_2O, which causes it to produce white fumes in the air. Water at 60° will absorb over 450 times its volume of the gas, producing the liquid known as "*Muriatic Acid.*" It dissolves the metals, and forms chlorides. When pure it is colorless, but has ordinarily a yellow tinge, due to various impurities. Its tests are H_3N, with which it forms a white cloud of sal-ammoniac fumes, and silver nitrate, from which it precipitates AgCl. With HNO_3 it makes aqua-regia,* or *royal water,* so called because it dissolves Au, the "king of the metals;" Cl is set free, which, in its nascent state, attacks the Au and combines with it.

***Boil HCl in a test tube with fragments of gold-leaf. They will not dissolve. Add a few drops of HNO_3, and a yellow solution of gold chloride will be quickly formed.**

Calcium Hypochlorite ($CaCl_2O_2$) is an ingredient of chloride of lime or bleaching powder. This is prepared by passing a current of Cl over pans of fresh slacked lime. It is much used in bleaching and as a disinfectant.

Calcium Chloride, the other compound of bleaching powder, was made in preparing CO_2 (see p. 74). It is used by chemists for drying gases. It absorbs H_2O so greedily that in the open air it will soon dissolve.

BROMINE—named from its bad odor—is a poisonous, volatile, deep-red liquid, with the general properties of Cl.* It is principally found in sea-water, forms bromides with the metals, and is used in photography and medicine.

FLUORINE is the only element that will not unite with O. It exists, in small quantities, in the enamel of the teeth. It is found in Derbyshire or fluor spar (CaF_2), of which beautiful ornaments are made. It unites with H, forming hydrofluoric acid (HF), noted for its corrosive action on glass.† (See *Appendix.*) This eats out the silica or sand from the glass, and is therefore used for etching labels on glass bottles and on shop windows.—*Example:* Powdered fluor spar is placed in a lead tray, and covered with dilute H_2SO_4. The heat of a lamp applied beneath, for a moment only, liberates the gas in white fumes very rapidly. The plate of glass is covered with wax, and the design to be etched is traced upon it with a sharp-pointed instrument. This is then laid over the tray, and the escaping gas soon etches the lines laid

* Br is the only element, except Hg, which is liquid at ordinary temperatures.

† So delicate is the test that by this means the presence of F has been detected in fossil teeth.

bare into an appearance like ground glass. A solution of HF in H_2O is often sold for this purpose. It is kept in lead or gutta-percha bottles, combines with H_2O with a hissing sound, like red-hot iron, and must be handled with care, as a minute drop even will sometimes produce an ulcer.

IODINE is named from its beautiful violet-colored vapor. It is made from kelp (the ashes of sea-weed), and is found in sea-water and in some mineral springs. It crystallizes in bluish-black scales, emits a smell resembling that of Cl, sublimes * slowly, and is deposited in crystals on the sides of the bottle in which it is kept. I is sparingly soluble in H_2O, but readily in ether or alcohol. It inflames spontaneously when in contact with phosphorus.† Its compounds with the metals, called the iodides, are remarkable for their variety and brilliancy of color. (See *Appendix.*) It stains cloth a yellowish tint, which may be removed by a solution of potassium iodide. Its test is starch, forming the blue iodide of starch.‡ I reveals the presence of this substance in potatoes, apples, etc. § It is much used in medicine to scatter scrofulous or cutaneous eruptions and swellings.

* A body is said to *sublime* when it rises as vapor and condenses in the solid form; when it condenses as a liquid it is said to *distil.*

† Place on a clean, white dish a few scales of iodine and a bit of phosphorus as large as a pea. They will soon combine, igniting the phosphorus and subliming a part of the iodine.

‡ Mix one or two drops of a solution of potassium iodide with a little dilute starch mucilage; no change of color will occur. Add a single drop of Cl water to the mixture; an immediate coloration will occur, owing to the combination of the Cl with the K, while I is set free, which acts upon the starch. Add a little more chlorine water; the color disappears, owing to the formation of chlorine iodide, which is without action on starch.

§ Pour a few drops of a solution of iodine in alcohol on a freshly-cut potato or apple. Blue specks will show the presence of starch.

BORON.

Symbol, B........Atomic Weight, 10.9.

Boron is found in nature in combination with O, as boracic acid. This is abundant in the volcanic districts of Tuscany.* Along the sides of the mountains, series

Fig. 46.

Preparing Boracic Acid.

of basins are excavated and filled with cold water from the neighboring springs. Into these basins the jets of steam, charged with boracic acid, are conducted. The H_2O absorbs the acid, and becomes itself heated to the boiling-point. It is then drawn off into the next lower basin. This process is continued until the bottom one is reached, when the solution runs into leaden pans heated by the steam from the earth ; here the H_2O is evaporated, and the boracic acid collected.

* Throughout an area of nearly thirty miles, is a wild, mountainous region, of terrible violence and confusion. The surface is ragged and blasted. Everywhere there issue from the ground jets of steam, filling the air with most offensive odors. The earth itself shakes beneath the feet, and frequently yields to the tread, engulfing man and beast. "The waters below are heard boiling with strange noises, and are seen breaking out upon the surface. Of old, it was regarded as the entrance to hell. The peasants pass by in terror, counting their beads and imploring the protection of the Virgin."

Borax (Na_2O, $2B_2O_3$, $10H_2O$) is a salt of this acid. It is a natural production, obtained by the drying of certain lakes in Thibet, and lately found in California. When dissolved in alcohol it gives a peculiar green tint to the flame. This is an easy test of the presence of this acid. Borax is employed in welding. It dissolves the oxide of the metal, and keeps the surface bright for soldering. It softens hard water by uniting with the soluble salts of lime or magnesia, and making insoluble ones which settle and form a thin sediment in the bottom of pitchers in which it is placed.*

SILICON.

Symbol, Si.....Atomic Weight, 28....Specific Gravity, 2.49.

Sources.—Silicon is found in combination with O as silica (silicic anhydride, SiO_2), commonly called silex or quartz. So abundant is this oxide that it probably comprises nearly one-half of the earth's crust. (See *Geology*, p. 40.) It forms beautiful crystals and some of the most precious gems. When pure, it is transparent and colorless, as in rock crystal. Jasper, amethyst, agate, chalcedony, blood-stone, chrysoprase, sardonyx, etc., are all common flint-stone or quartz, colored with some metallic oxide. The opal is only SiO_2 and H_2O. Sand is mainly fine quartz, which, when hardened and cemented, we call sandstone. Yellow or red sand is colored by iron-rust.

Properties.—It is tasteless, odorless, and colorless. It

* Borax is also extensively used in "blow-pipe analysis." When it is melted with chromium oxide, it gives an emerald green; with cobalt oxide, a deep blue; with copper oxide, a pale green; with manganese oxide, a violet.

seems very strange to call such an inert substance an anhydride; yet it readily unites with the alkalies, neutralizes their properties, and forms a large class of salts known as the silicates, which are found in the most common rocks.—*Example:* feldspar, found in granite.

Silica in Soil and Plants.—Silica is insoluble in H_2O, unless it contains some alkali. When the silicates, so abundant in rocks, disintegrate and form soil, the alkali and silica are both dissolved in the water, and taken up by the roots of plants. We see the silex on the surface of scouring-rushes and sword-grass, which cut the fingers if handled carelessly. It gives stiffness to the stalks of wheat and other grains, and produces the hard, shiny surface of bamboo, corn, etc.

Petrifaction.—Certain springs contain large quantities of some alkaline carbonate; their waters, therefore, dissolve silica abundantly. If we place a bit of wood in them, as fast as it decays, particles of silica will take its place—atom by atom—and thus petrify the wood. The wood has not been *changed to* stone, but has been *replaced by* stone.

Compounds.—***The Silicates.***—Glass* is a mixture

* Glass was known to the ancients. Hieroglyphics, that are older than the sojourn of the Israelites in Egypt, represent glass-blowers at work, much after the fashion of the present. In the ruins of Nineveh, articles of glass, such as vases, bowls, etc., have been discovered. Mummies, three thousand years old, are adorned with glass beads. The inventor is not known. Pliny tells us that some merchants, once encamping on the sea-shore, found in the remains of their fire bits of glass, formed from the sand and ashes of the sea-weed by the heat; but this is impossible, as an open fire is not sufficient to melt these materials. In the fourth century B.C., the glass-works at Alexandria produced exquisite ornaments, with raised figures beautifully cut and gilded. But in the twelfth century A.D., glass was still so costly in England that glass windows were thought very magnificent; and, as late even as about 1500, when the great Earl of Northumberland left one of his houses for a time, he was careful to have the glass of the windows taken down and packed for safe keeping.

of several silicates. There are four varieties used in the arts. 1. *Window or plate glass* is composed of silicates of calcium and sodium. It is made by heating white sand, sal-soda, and lime in clay crucibles for about forty-eight hours, when the materials fuse and combine into a double silicate. The Ca hardens and gives lustre; the Na renders the glass fusible, but imparts a green tint. 2. *Bohemian glass* consists of silicates of calcium and potassium. Unlike Na, K gives no color. 3. *Flint-glass* * or *crystal* contains silicates of potassium and lead. The latter is used in large quantities and produces a soft, lustrous glass, which can be ground into imitation gems, table-ware, chandelier pendants, prisms, etc. 4. *Green bottle-glass* is made of silicates of calcium, sodium, aluminum, and iron. The last gives the opaque green of the common junk bottle.

Coloring Glass.—A small quantity of some metallic oxide melted with the glass furnishes any tint desired: Co gives a beautiful sapphire blue; Au or Cu, a ruby-red; Mg, a violet; U, a yellow; As, a soft white enamel, as in lamp-shades; and Sn, a hard enamel, as in watch-faces.

Annealing Glass.—If the glass utensils were used immediately, they would be found extremely brittle, and would drop in pieces in the most unaccountable way. The heat of the hand or a draft of cool air would sometimes crack off the thick bottom of a tumbler. They are therefore cooled very gradually for days, which allows the particles to assume their natural place, and the molecular attractions to become equalized. †

* So called because pulverized flint was formerly used for sand.

† This principle is beautifully illustrated by the philosophical toy known as the "Prince Rupert's Drop." (See *Physics*, p. 46.)

Ornamental Ware.—Venetian balls or paper weights are made by arranging bits of colored glass in the form of fruits, flowers, etc., and then, inserting them in a hollow globe of transparent glass, still hot, the workman draws in his breath, and the pressure of the air above collapses the globe upon the colored glass, and leaves a concave surface in the opposite side of the weight. The lens form always magnifies the size of the figures within.

Tubes and Beads.—In making glass tubing, the workman inserts his iron blowing-tube in a pot of melted glass, and gathers upon the end a suitable amount; drawing this out, he blows into the tube, swelling the glass into a globular form. Another dip into the pot and another blow increase its size, until at last a second workman attaches an iron rod to the other end. The two men then separate at a rapid pace. The soft glass globe diminishes in size as it lengthens, until at last it hangs between them a glass tube of a hundred feet in length, and perhaps only a quarter of an inch in diameter.

For making beads, glass tubes are cut in short pieces, and then worked about in a mixture of wet ashes and sand, until they are filled. They are next put with loose sand in a cylinder rapidly revolving over a hot furnace. The heat softens the glass, but the mixture within presses out the sides, and the sand grinds the edges, until at last the beads become round and perfect, and are taken out ready for market.

SULPHUR.

Symbol, S. . . . Atomic Weight, 32. . . . Specific Gravity, 2.

Sources.—S is found native in volcanic regions. It is mined at Mount Ætna in great quantities. United with the metals it forms sulphides, known as cinnabar, iron pyrites, galena, blende, etc. Combined with O it exists in gypsum (plaster), heavy spar, and other sulphates. It is found in the hair, and many dyes contain Pb which unites with the S, and forms a black compound that stains the hair. It is contained in eggs, and so tarnishes our spoons by forming a sulphide of silver. It is always present in the flesh, and hence manifests itself in our perspiration. In commerce it is sold as brimstone, formed by melting S and running it into moulds; also as flowers of sulphur, obtained by sublimation.

Properties.—It is insoluble in H_2O, and hence tasteless. Its solvents are carbon disulphide (CS_2), oil of turpentine, and benzole. It is a non-conductor of heat, and crackles when we grasp it with a warm hand. It manifests itself under four allotropic forms: 1st, octahedral crystals; 2d, prismatic crystals; 3d, an amorphous (without form) or uncrystallized state; and 4th, a viscid condition. The last is the most interesting.—*Example:* When S is melted, and then heated more strongly, it changes to a thick, viscid, dark-colored liquid resembling molasses. If this is poured into cold water, it becomes elastic like india-rubber. In this form it is used for taking impressions of medals, coins, etc. (See *Appendix.*)

Uses.—On account of its ready inflammability, S is employed in the making of matches and gunpowder, but its chief consumption is in the production of H_2SO_4.

Compounds.—*Sulphurous Anhydride,* SO_2, an irrespirable, suffocating gas, is formed by S burning in the air, as in the lighting of a match. It is very poisonous, and extinguishes combustion. If our "chimney burns" at any time, we can easily quench the flame by pouring a little S into the stove.

Uses.—SO_2 is used for bleaching silk, straw, and woollen fabrics. Cl turns them yellow, but SO_2 unites with the coloring matter, and forms a colorless compound. Its action is therefore very different from that of Cl.—*Example:* A red rose, bleached in the fumes of burning S, can be restored to its original color by very dilute H_2SO_4. This acid being stronger, neutralizes the action of the SO_2. New flannels, washed in strong soap, turn yellow, because the alkali of the soap unites with the SO_2 used in bleaching the cloth, and thus sets free the original color. S is also frequently employed to check fermentation, as when it is burned in a barrel before filling with new cider.

Sulphuric Anhydride, SO_3, may be prepared by the oxidation of SO_2 or by removing H_2O from H_2SO_4. It is often called *anhydrous sulphuric* acid. If Nordhausen acid * be heated, the vapors may be condensed in a mass of silky, crystalline fibres of SO_3. This will show no acid reaction, will not redden blue litmus-paper, and, if the fingers are dry, can be molded like wax. If it be dropped into H_2O, it will hiss like a red-hot iron, and

* So named from the German town near which it was formerly made by the distillation of green vitriol (iron sulphate).

forming H_2SO_4, will exhibit all the properties of that corrosive substance.

Sulphuric Acid, *Oil of Vitriol,* is the king of the acids. It is of the utmost importance to the manufacturer and chemist, as it is used in the preparation of nearly all other acids, and forms many valuable compounds.

Preparation.—If we burn a little S in a bottle it will soon become filled with a white cloud of SO_2. Now another atom of O would make this SO_3, sulphuric anhydride. Nitric acid, it will be remembered, easily parts with its O. So if we stir the SO_2 with a swab wet in aqua-fortis, we shall quickly see the familiar hyponitric acid fumes, indicating that the acid has been decomposed and has given up its O. Add a little water and shake the jar thoroughly. On testing the liquid with a few drops of a solution of barium chloride, the beautiful white precipitate will prove the presence of H_2SO_4.*

Fig. 47.

Making H_2SO_4.

The Manufacture of Sulphuric Acid on a large scale is based on the principle of the preceding illustration. The process is facilitated by the curious fact that the nitric oxide (NO) produced by the decomposition of the HNO_3 has the property of acting as a carrier of O between the common air and SO_2, whereby it can oxidize an almost indefinite quantity, thus forming SO_3, which, in the presence of H_2O, will be at once converted into H_2SO_4. S is burned in a current of air in furnaces A, A. In the stream of heated gas is suspended an

* The reaction in making the acid may be thus expressed: $2HNO_3 + SO_2 = H_2SO_4 + 2NO_2$.

iron pot, *b*, charged with a mixture of sodium nitrate and H_2SO_4. Vapors of HNO_3 are thus set free, and these pass on mixed with SO_2 and excess of atmospheric air.

Fig. 48.

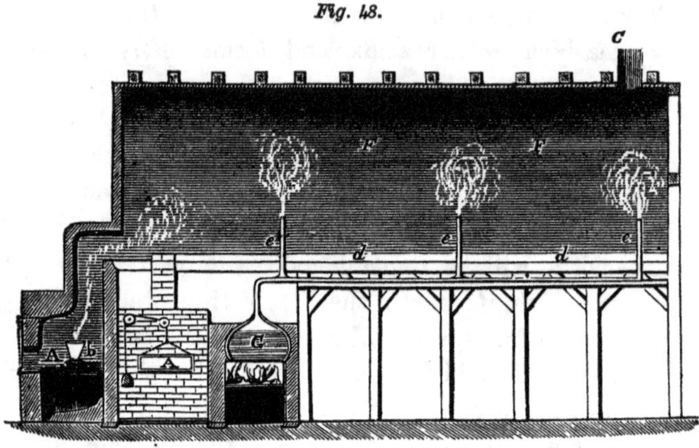

Manufacture of H_2SO_4.

The mingled gases pass into immense chambers, F, of sheet lead. A shallow layer of H_2O, *d*, covers the floor, and the intermixture and chemical action of the gases are further favored by the injection of jets of steam, *e*, supplied from the boiler, G.

The chemical action which ensues may be explained as follows:—The nitric acid is quickly reduced to nitric oxide, NO. This takes up an atom of O from the air, becoming NO_2, and flies back to the SO_2 making a molecule of SO_3, which, with a molecule of H_2O becomes H_2SO_4, a molecule of sulphuric acid. The NO once more seeks the air and returns laden with O for the SO_2. This process continues until the chamber becomes so full of the sluggish N that the other gases are nearly lost in it, when they are allowed to escape gradually. The weak sulphuric acid which collects on the floor is drawn off and condensed by evaporation in lead pans, and finally, when it begins to corrode the lead, in platinum or glass stills. It is lastly put in large bottles packed in boxes called *carboys*, when it is ready for transportation.

Properties.—It is a dense, oily liquid, without odor, and of a brownish color. Its affinity for moisture is most remarkable. If exposed in an open bottle it gradually absorbs water from the air, and increases in bulk, sometimes even doubling its weight. It blackens wood and other organic substances, by taking away their H_2O and leaving the C.* When mixed with H_2O, it occupies less space than before, and produces much heat; 4 parts of acid to 1 of H_2O will boil a test-tube of water. It commonly contains lead, which falls as a milky precipitate ($PbSO_4$) when the acid is diluted. It is the strongest of the acids, and will displace the others from their compounds. It stains cloth red, but the color can be restored by an alkali, if applied immediately. Its test is barium chloride, which forms a white, cloudy precipitate. In this way a drop of H_2SO_4 can be detected in a quart of H_2O.

Hydrogen Sulphide, H_2S, *Sulphuretted Hydrogen, Sulphydric Acid.*—This gas is produced in the decay of organic matter, and is always found near cess-pools, drains, and sinks, turning lead paint black and emitting a disagreeable smell. It gives the characteristic odor to the

* Strong oil of vitriol poured on a little loaf-sugar moistened with hot water, will cause an energetic boiling and a copious formation of black charcoal. Sugar consists of water and charcoal, and gives up the former to satisfy the acid.

Fig. 49.

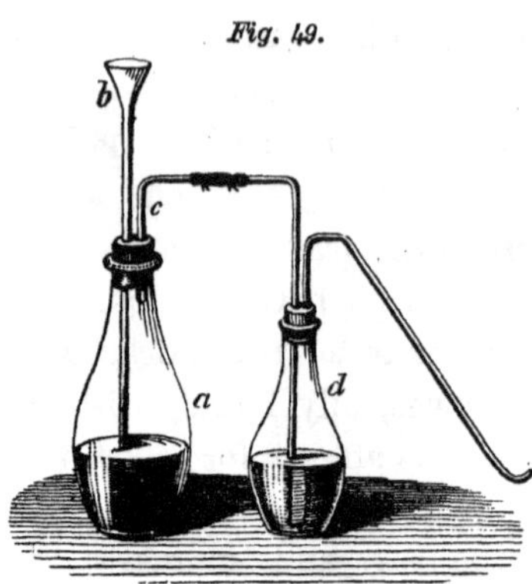

Preparing H_2S.

mineral waters of Avon, Clifton, Sharon, and other celebrated sulphur springs. It is prepared by the action of dilute H_2SO_4 upon ferrous sulphide (FeS). The reaction is as follows: $FeS + H_2SO_4 = FeSO_4 + H_2S$.

H_2S has the disgusting odor of rotten eggs. It is very poisonous, and therefore makes an open sewer destructive to health. Its solution in H_2O is much used in the laboratory to precipitate many of the metals as sulphides. Its test is lead acetate (sugar of lead.)

Carbon Disulphide, CS_2, is produced by passing the vapor of S over red-hot coals. It is a volatile, colorless liquid, and has never been frozen. The fact that a yellow, odorless solid thus unites with a black, odorless solid to form such a colorless, odoriferous liquid, illustrates very finely the power of chemical affinity. CS_2 readily dissolves S, P, and I. It is a powerful refractor of light, and is used for filling hollow, glass prisms employed in experiments with the solar spectrum. In its combustion it unites with O, forming CO_2 and SO_2.

PRACTICAL QUESTIONS.

1. If chlorine water stands in the sunlight for a time, it will only redden a litmus-solution. Why does it not bleach it?
2. Why do tinsmiths moisten with HCl, or sal-ammoniac, the surface of metals to be soldered?

3. How much HCl can be made from 25 lbs. of common salt?

4. What weight of NaCl would be required to form 25 lbs. of muriatic acid?

5. HCl of a specific gravity of 1.2 contains about 40 per cent. of the gas. This is very strong commercial acid. What weight could be formed by the HCl acid gas produced in the reaction named in the preceding problem?

6. What is the difference between sublimation and distillation?

7. Why do eggs discolor silver spoons?

8. Explain the principle of hair-dyes.

9. Why is new flannel apt to turn yellow when washed?

10. Is it safe to mix oil of vitriol and water in a glass bottle?

11. What is the color of a sulphuric acid stain on cloth? How would you remove it?

12. What causes the milky look when oil of vitriol and water are mixed?

13. What is the chemical relation between animals and plants? Which perform the office of reduction, and which of oxidation?

14. How many pounds of S are contained in a cwt. of H_2SO_4?

15. How much O and H_2O are needed to change a ton of SO_2 to H_2SO_4?

16. How much O in a lb. of H_2SO_4?

17. State the analogy between the compounds of O and S.

PHOSPHORUS.

Symbol, P....Atomic Weight, 31....Specific Gravity, 1.83.

THE name Phosphorus signifies *light-bearer*, given because this substance glows in the dark. It was called by the old alchemists "the son of Satan."*

* The following singular event is said to have occurred many years before the reputed discovery of phosphorus by Brandt in 1669. A certain Prince San Severo, at Naples, exposed some human skulls to the action of several reagents, and then to the heat of a furnace. From the product he obtained a substance which burned for months without apparent loss of weight. San Severo refused to

Sources.—It exists in small quantities in rocks, and by their decay passes into the soil, is taken up by plants, is then stored in their seeds (wheat, corn, oats, etc.), and finally passes into our bodies. As calcium phosphate ("phosphate of lime"), it is a prominent constituent of our bones.* Phosphorus is so necessary to the operation of the brain that the alchemists had a saying, "No phosphorus, no brains."

Fig. 50.

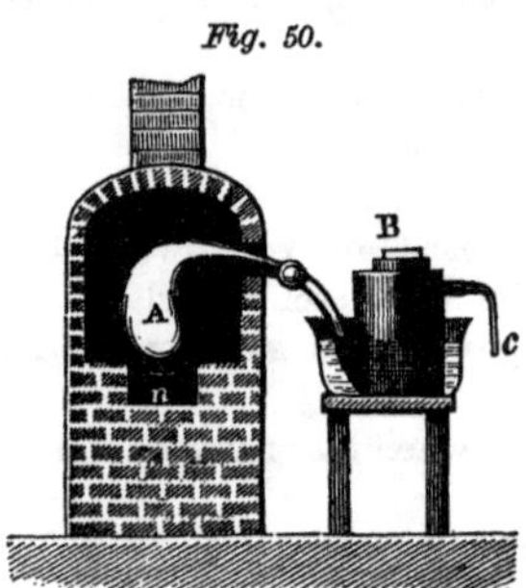

Manufacture of Phosphorus.

Preparation.—It is prepared in immense quantities from bones. These are first calcined to whiteness to burn out the animal matter, then treated with H_2SO_4 to remove the Ca (pp. 140, 230), and lastly heated to a high temperature with C to deoxidize the phosphorus, which distils as a vapor, and is condensed under H_2O.

Properties.—It is a waxy, translucent solid, at all temperatures above 32° emits a feeble light, melts at 111°, and ignites at a little higher temperature. It should be handled with the utmost care, always kept and cut under H_2O, and never used except in very small quantities. Its burns are deep and dangerous. It is poisonous, and its vapor produces horrible ulcerations of the jawbone in workmen who use it.

Amorphous Form.—Heated for several hours at

divulge the process, as he wished his family vault to be the only one to possess a "*perpetual lamp*," the secret of which he considered himself to have discovered.

* "Of phosphorus every adult person carries enough (1¾ lbs.) about with him in his body to make at least 4,000 of the ordinary two-cent packages of friction matches, but he does not have quite sulphur enough to complete that quantity of the little incendiary combustibles."—NICHOLS's *Fireside Science.*

a temperature of about 500°, in a close vessel filled with N or CO_2, the melted phosphorus changes into a brick-red solid, and seems to lose all its former properties. It is now insoluble in CS_2, which can be used to dissolve out every trace of the common form. Its specific gravity is increased to 2.14. It can be handled with impunity, carried in the pocket like so much snuff, and even heated to nearly 400° without taking fire. At a little over 500°, however, it changes into the common form and bursts into a blaze.

Uses.—*Matches.*—The principal use of phosphorus is in the manufacture of matches. 1. *The Lucifer Match.*—The bits of wood are first dipped in melted S and dried; then in a paste of phosphorus, nitre, and glue, which completes the process. The object of the nitre is to furnish O to quicken the combustion. Instead of this, potassium chlorate is sometimes used; it can be recognized by a crackling sound and jets of flame when ignited. The tips are colored by red-lead, or Prussian blue, mixed in the paste. When a match is burned, the reaction is as follows: first, the friction ignites the phosphorus, which burns, forming P_2O_5; * this produces heat enough to inflame the S, which makes SO_2; lastly, the wood takes fire, and forms CO_2 and H_2O. Thus there are four compounds produced in the burning of a single match.

2. *The Safety Match.*—The pieces of wood are dipped into melted paraffine (see p. 205) and dried. They are

* The burning phosphorus produces a very luminous flame, because of the reflection of light from the dense vapor (P_2O_5). The following experiment is very suggestive in this connection: Ignite a bit of phosphorus placed upon a sheet of white paper. The paper will be blackened just where the phosphorus lay, but will not take fire; and after the flame is extinguished, one can write upon it with pen and ink, close to the edge of the charred portion.

then capped with a paste of potassium chlorate, sulphide of antimony, powdered glass, and gum-water. They ignite only when rubbed on a surface covered with a mixture of red phosphorus and powdered glass.

Phosphorescence.—The luminous appearance of putrefying fish and decayed wood is well known. The latter is sometimes called "fox-fire." The "glow-worm's fitful light" is associated with our memory of beautiful summer evenings. In the West Indies, fire-flies are found that emit a green light when resting, and a red one when flying. They are so brilliant that one will furnish light enough for reading. The natives wear them for ornaments on their bonnets, and illuminate their houses by suspending them as lamps. The ocean occasionally takes on strange colors, and the sailor finds his vessel plowing at one time apparently a furrow of fire, and at another one of liquid gold. The water is all aglow, and the flames seem to leap and dance with the waves or the motion of the ship. The phenomenon is produced by multitudes of animalcules which frequent certain seas.*

Compounds.—*Hydrogen Phosphide,* H_3P, *Phosphuretted Hydrogen,* is formed in the decomposition of bones and organic substances. It is a poisonous gas, remarkable for its disgusting odor, for igniting spontaneously on coming to the air, and for the singular beauty of the rings formed by its smoke. It is prepared by heating in a retort a strong solution of potash containing a few bits of phosphorus. It has been thought by some that the Will-o'-the-wisp, Jack-o'-the-lantern, etc., as seen near graveyards and in swampy places, are produced by

* Many substances, after having been exposed to the light, will shine for some time when removed into the darkness. They possess the property of absorbing invisible actinic-rays and slowly emitting them as visible light-rays. Thus, a dial coated with the so-called *Luminous Paint* (calcium sulphide) will show the time at night

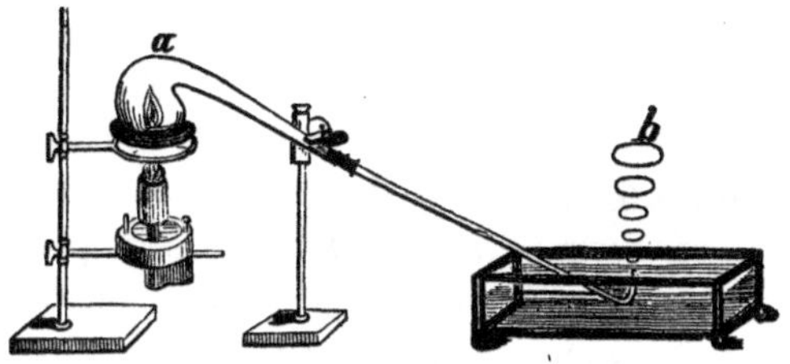

Preparation of H_3P.

this gas coming off from decaying substances, and igniting as it reaches the air.

ARSENIC.

Symbol, As. . . . Atomic Weight, 75. . . . Specific Gravity, 5.9.

Volatilizes without fusion at about 356° F.

As is a brittle, steel-gray metal,* commonly sold, when impure, as *cobalt*.† If heated in the open air it gives off the odor of garlic, which is a test of As.

* Arsenic very much resembles phosphorus in its general properties, and is therefore classified with it, but it conducts electricity moderately, and has a high brilliancy. It seems to be intermediate between the non-metals and the metals.

† Cobalt is a reddish-white metal, found in combination with arsenic. Co received its name from the miners, because its ore looked so bright that they thought they would obtain something valuable; but when, by roasting, it crumbled to ashes, they believed themselves mocked by the evil spirit (Kobolt) of the mines. The oxide of cobalt makes a beautiful blue glass, which, when ground fine, is called *smalt*. It is used for tinting paper, and by laundry women to give the finished look to cambrics, linen, etc. Its impure oxide, called *zaffer*, imparts the blue color to common earthenware and porcelain. The chloride ($CoCl_2$) is used as a sympathetic ink. Letters written with a dilute solution of it are invisible when moist with the H_2O absorbed from the air, but on being dried at the stove, again become blue. If the paper be laid aside the writing will disappear, but may be revived in the same manner. A winter landscape may be drawn with India-ink, the leaves being added with this ink. On being brought to the fire it will bloom into the foliage of summer.

Arsenious Anhydride, As_2O_3.—This is the well-known "*ratsbane,*" and is sometimes sold as simply "arsenic."

Preparation.—It is made in Silesia, by roasting arsenical iron-ore at the bottom of a tower, above which is a series of rooms through which the vapors ascend, and pass out at a chimney in the top. The As burns, forming As_2O_3, which collects as a white powder on the walls and floors of the chambers above.*

Properties.—"Arsenic" is soluble in hot H_2O, and has a slightly sweetish taste. It is a powerful poison, doses of two or three grains being fatal, although an over-dose acts as an emetic. It is an antiseptic, and so in cases of poisoning frequently attracts attention by the preservation of parts of the body, even twenty or thirty years after the murder has been committed. The antidote is milk or whites of eggs.†

Marsh's Test.—There is no other poison which is so easily detected. Prepare a flask for the evolution of H. Ignite the jet of gas, and hold in the flame a cold porcelain dish. If it remains untarnished, the materials contain no As. Now pour in through the funnel-tube a few drops of a solution of As;‡ the color of the flame will be seen to change almost instantly, and a copious "metallic mirror" of As will be deposited on the dish.

* Its removal is a work of great danger. The workmen are entirely enveloped in a leathern dress and a mask with glass eyes; they breathe through a moistened sponge, thus filtering the air of the fine particles of arsenic floating through it. Yet, in spite of all these precautions, they rarely live beyond forty.

† The exact chemical antidote is hydrated ferric oxide. In this, as in most other cases of poisoning, where the antidote is not at hand, an emetic should be taken at once—a tea-spoonful of mustard in a glass of warm water, or even a quantity of soap-suds. (See *Physiology*, page 209.)

‡ This is made by dissolving a little As_2O_3 in HCl.

The gas formed in this experiment—arseniuretted hydrogen—is very poisonous indeed, and the utmost care should be used to prevent its inhalation.*

Fig. 52.

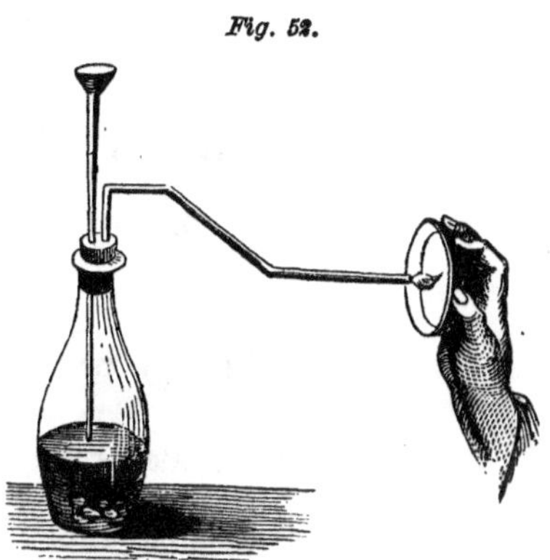

Marsh's Test.

Arsenic-Eating.—It is said that the peasants in a portion of Hungary are accustomed to eat As, both fasting and as a seasoning to their food. A very minute portion will warm, stimulate, and aid in climbing lofty mountains. The arsenic-eaters are described as plump and rosy, and it is said that the young people resort to this dangerous substance, as a species of cosmetic. They begin with small doses, which are gradually increased; but if the person ceases the practice, all the symptoms of arsenic poisoning immediately appear. Horse-jockeys sometimes feed arsenic to their horses to improve their flesh and speed.

* In a case of poisoning, of course, the contents of the stomach would be substituted for the solution of As, and other tests besides this would be employed. We can imagine with what care a chemist would conduct the examination, and with what intense anxiety he would watch the porcelain dish as the flame played upon it, hesitating, and dreading the issue, as he felt the life of a fellow-being trembling on the result of his experiment.

THE METALS.

THE METALS OF THE ALKALIES.

K, Na, L, Cs, Rb, and H_4N(?).

POTASSIUM.

Symbol, K....Atomic Weight, 39....Specific Gravity, 0.86.

Source.—K is found abundantly in the various rocks, which by their decomposition furnish it to the plants from which we obtain our entire supply.*

Preparation.—This metal was discovered by Sir Humphrey Davy, in 1807. On passing the current of a powerful galvanic battery through potash, the globules of the K appeared at the negative pole. The metals Na, Ba, Sr, and Ca, were afterward separated in the same manner. This discovery constituted a most important epoch in chemistry. K is now prepared by distilling in iron bottles, at an intense heat, potassium carbonate and charcoal. The green vapors of K are condensed in receivers of naphtha, and CO passes off as a gas. $K_2CO_3+2C=K_2+3CO$. It is a difficult and dangerous process. The vapor

* "An acre of wheat producing twenty-five bushels of grain and 3,000 lbs. of straw, removes about 40 lbs. of potash in the crop. An acre of corn, producing 100 bushels, removes in kernel and stalk 150 lbs. of potash and 80 lbs. of phosphoric acid. An acre of potatoes, yielding 300 bushels, will remove in tubers and tops 400 lbs. of potash and 150 lbs. of phosphoric acid. A pound of wheat holds a quarter of an ounce of mineral substances, and a pound of potatoes one-eighth of an ounce."—NICHOLS.

takes fire instantly on contact with air or water. It also absorbs CO, and the compound, if kept, becomes powerfully explosive. To prevent this danger, the K is immediately redistilled.

Properties.—K is a silvery-white metal, soft enough to be spread with a knife, and light enough to float like cork. Its affinity for O is so great that it is always kept under the surface of naphtha, which contains no O. K, when thrown on H_2O, decomposes it, sets free one atom of H, and forms KHO. The heat developed is so great, that the H catches fire and burns with some volatilized K, which tinges the flame with a beautiful purple tint. If the H_2O be first colored with red litmus, it will become blue by the alkali formed.

Fig. 53.

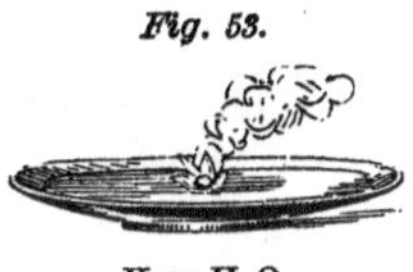

K *on* H_2O.

Compounds.—Potash, K_2O, has so great an affinity for H_2O that the anhydrous form is rarely prepared. Its hydrate, KHO,† is a white solid made from potassium carbonate by the action of slacked lime. It is the most powerful alkali. It neutralizes the acids, and turns red litmus to blue. It is used to cauterize the flesh, and is hence commonly called "caustic potash." It dissolves the cuticle of the finger which touches it, and so has an unctuous feel. It unites with grease, forming soap, in the manufacture of which it is extensively used.

Potassium Carbonate, K_2CO_3,‡ *Pearlash, "Carbonate of Potash,"* is obtained in the following manner:

* Cut the metal in small pieces and cover it with a receiver, since the melted globule bursts at the close of the experiment.

† $K_2O+H_2O=2(KHO)$, or 2 molecules of potassium hydrate.

‡ The symbol K_2CO_3 is merely a list of the elements, and the proportion of each contained in a molecule of potassium carbonate. It is called an *empirical*

Potash exists in plants, combined with various acids, such as tartaric, malic, oxalic, etc. When the wood is burned, the organic acids are decomposed by the heat, and the CO_2 combines with the K_2O, forming K_2CO_3. The ashes are then leached and the *lye* is evaporated, when the salt crystallizes. This forms potassium carbonate, the "potash" of commerce. When refined it is called "pearlash." Where wood is abundant, immense quantities are burned solely for this product. Birch gives the purest potash, while the leaves of a tree furnish twenty-five times as much as the heart.*

Hydrogen Potassium Carbonate,† $HKCO_3$, *Saleratus.* "*Bicarbonate of Potash,*"‡ is prepared by pass-

formula. It can be written thus: $K_2O.CO_2$, and is then termed a *rational formula*, since it indicates the compounds which, put together, form the carbonate. One objection to the latter formula is that we do not know that the separate compounds still exist in the salt. The empirical formula contains all that is positively decided.

* Vast deposits of potash have been opened up to us at the Stassfurth salt mines in Germany, the supply from which is more than from the wood-ash sources of the whole world. "Only about 13,000 tons of potash were sent to market from the United States and British America in 1870, and yet from Stassfurth, where a dozen years ago it was not supposed that a single ton could be produced, 30,000 tons of potassium chloride were manufactured and supplied to consumers upon both continents during the following year. The surface salts at these mines, which hold the potash, are practically inexhaustible, and millions of tons will be supplied in succeeding years."—*Fireside Science.*

† The molecule of carbonic acid is H_2CO_3. In potassium carbonate, K_2CO_3, both the atoms of H contained in the carbonic acid are replaced by the metal K; in hydrogen potassium carbonate, $HKCO_3$, only one atom of H is thus replaced. In this way two classes of salts are derived; the so-called *acid salts*, where only one atom of H has been replaced, and the *neutral salts*, where both atoms have been replaced by a metal. Thus hydrogen potassium sulphite, $HKSO_3$, is an acid salt, and potassium sulphite, K_2SO_3, is a neutral salt. An acid containing two atoms of H, capable of displacement by a metal, is said to be *dibasic*, as H_2SO_4, H_2CO_3; and one having three atoms capable of displacement is termed *tribasic*. *Example:* H_3PO_4, which forms three different salts from Na, the H being displaced from the acid step by step.

‡ If we double the number of atoms of each element in a molecule of hydrogen potassium carbonate, the rational formula will be $K_2O.H_2O.2CO_2$; hence this salt is commonly called the bicarbonate of potash.

ing CO_2 through a strong solution of potassium carbonate.

Potassium Nitrate, KNO_3,* *Nitrate of Potash, Saltpetre, Nitre.*—This salt is found as an efflorescence on the soil in tropical regions, especially in India. It is obtained thence by leaching.† It is formed artificially by piling up great heaps of mortar, refuse of sinks, stables, etc. "In about three years, these are washed, and each cubic foot of the mixture will furnish four or five ounces of saltpetre." It dissolves in about three and a half times its weight of cold H_2O.

Properties and Uses.—It is cooling and antiseptic; hence it is used with common salt ($NaCl$) for preserving meat. It parts readily with its O, of which it contains nearly 48 per cent., and deflagrates brilliantly. Every government keeps a large supply on hand for making gunpowder, in the event of war. Gunpowder is now composed of about six parts, by weight, of nitre, and one each of charcoal and sulphur—the proportion varying with the purpose for and the country in which it is made. Its explosive force is due to the expansive power of the gases formed. At the touch of a spark the saltpetre gives up its O to burn the S and C. The reaction that ensues may be approximately represented as follows: $2KNO_3+S+3C=K_2S+N_2+3CO_2$.

N and CO_2 are gases, and in the great heat of perhaps 2,000°, high enough to melt silver or copper, the K_2S becomes a vapor. With the sudden increase of temperature they all expand till they occupy at least 1,500 times the

* In this salt the H of HNO_3 is replaced by K. (See page 23, note.)

† It was manufactured in the Mammoth Cave, Kentucky, during the war of 1812. The remains of the works, and even the deep ruts of the wagon-wheels, are still to be seen, preserved in the pure, still air.

space of the powder.—MILLER. The bad odor of burnt powder is due to the slow formation of H_2S in the residuum. Fireworks are composed of gunpowder ground with additional C and S, and some coloring matter. Zinc filings produce green stars, steel filings variegated ones. $Sr2NO_3$ tinges flame with crimson. Salts of copper give a blue or green light, and camphor a pure white one.

Potassium Chlorate, $KClO_3$, is a white, crystallized salt much used in preparing oxygen, making matches, fireworks, etc. It is a powerful oxydizing agent.*

Potassium Bichromate † is a red salt highly valued in dyeing, calico-printing, and photo-lithography. If we mix a solution of this salt and one of sugar of lead, a yellow-colored precipitate will be formed, known in the arts as chrome-yellow (lead chromate).

SODIUM.

Symbol, Na.... Atomic Weight, 23.... Specific Gravity, 0.972.

THIS metal is found principally in common salt. Its preparation is similar to that of K, but is more easily managed. It is very like K in appearance, properties,

* *Examples:* 1. Cover a bit of phosphorus, no larger than a mustard seed, with finely powdered $KClO_3$ (See *Appendix*), wrap in a paper and lay it on an anvil. Upon striking the mixture with a hammer, a sharp detonation will ensue. 2. Place in a wine-glass five or six pieces of phosphorus as large as a grain of wheat, and cover with crystals of $KClO_3$. Fill the glass two-thirds full of H_2O. By means of a pipette, or a glass funnel, introduce into immediate contact with the $KClO_3$ a few drops of strong H_2SO_4. A violent chemical action will immediately ensue, and the phosphorus will burn under the water with vivid flashes of light.

† Chromic anhydride (Cr_2O_3) is an oxide of chromium (*chroma*, color), a metal prized only for its numerous brilliantly colored compounds. It is rather rare, and mainly found in chrome iron-stone ($FeO.Cr_2O_3$).

and reaction. When thrown on H_2O it rolls over its surface like a tiny silver ball; if the H_2O be heated, it bursts into a bright yellow blaze. The test of all the soda salts is the yellow tint which their solution in alcohol gives to flame.

Compounds.—*Sodium Chloride,* NaCl, *Common Salt,* is a mineral substance absolutely necessary to the life of human beings and the higher orders of animals. It does not enter into the composition of tissue, but is essential to the proper digestion of the food and to the removal of worn-out matter. (See *Physiology,* p. 137.) Among the many cruel punishments inflicted in China, deprivation of salt is said to be one, causing at first a most indescribable longing and anxiety, and finally a painful death. As salt is so universally necessary, it is found everywhere. Our Father, in fitting up a home for us, did not forget to provide for all our wants. The quantity of salt in the ocean is said to be equal to five times the mass of the Alps. Salt lakes are scattered here and there; saline springs abound; and besides these, in the earth are stored great mines, probably produced by the evaporation of salt lakes in some ancient period of the earth's history. Near Cracow, Poland, is a bed five hundred miles long, twenty miles wide, and a quarter of a mile thick. In Spain, and lately in Idaho, it has been quarried out in perfect cubes, transparent as glass, so that a person can read through a large block.

Preparation.—On the sea-shore it is manufactured by the evaporation of sea-water, each gallon containing about four ounces.* At Syracuse, New York, near by

* Salt is soluble in less than three times its weight of H_2O. It is scarcely more soluble in hot than cold H_2O, and a *saturated* solution (one containing all it

and underneath the Onondaga Lake, is apparently a great basin of salt-water, separated from the fresh-water above by an impervious bed of clay. Upon boring through this, the saline water is pumped up in immense quantities. The H_2O is evaporated by heating in large iron kettles over a fire, or in shallow, wooden vats by exposure to the sun—whence the name "solar salt." If boiled down rapidly, fine table-salt is made; if more slowly, coarse salt, as large crystals have time to form. Frequently they assume a "hopper shape," one cube

Fig. 54.

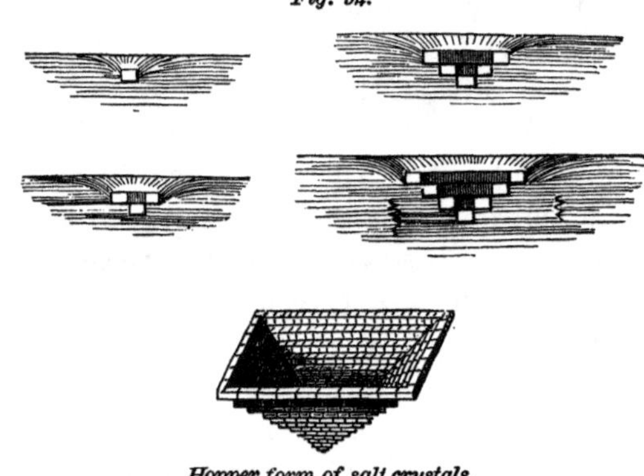

Hopper form of salt crystals.

appearing, then others collecting at its edges, and gradually settling, until a hollow pyramid of salt-cubes, with its apex downward, is formed.

Uses.—NaCl is used largely as a fertilizer, for preserving meats and fish, and for preparing Cl, HCl, and the various compounds of Na.

will dissolve) has about 36 per cent. Sea-water contains about 3 per cent. Sodium carbonate was formerly obtained from the ashes of sea-plants, as potassium carbonate is now from the ashes of land-plants.—ROSCOE.

Sodium Sulphate (Na_2SO_4, $10H_2O$), *Glauber's Salt*, named from its discoverer, is made in great quantities from NaCl, as the first stage in the manufacture of sodium carbonate. It is remarkably efflorescent, the salt, by exposure to the air, losing its ten molecules of H_2O.* It has a bitter, saline taste and is used in medicine.

Sodium Carbonate (Na_2CO_3, $10H_2O$), *Sal-soda*, is used extensively in the arts. It is, therefore, of great importance to all consumers of soap, glass, etc., that it should be manufactured as cheaply as possible. Leblanc's process of making it from NaCl is now generally adopted. The operation comprises two stages: Changing, 1. NaCl into Na_2SO_4; and, 2. Na_2SO_4 into Na_2CO_3.

1. A mixture of NaCl and H_2SO_4 is heated. Na_2SO_4 is formed with a copious evolution of HCl. The fumes of this gas are conducted into the bottom of a vertical flue filled with pieces of coke wet with constantly falling H_2O. The gas is here absorbed and a weak muriatic acid formed in great quantities.†

2. The Na_2SO_4 is heated with chalk ($CaCO_3$) and charcoal. The C deoxidizes the Na_2SO_4, changing it into Na_2S. The metals of the Na_2S and the $CaCO_3$ change places,

* *Experiment:* Make a saturated solution of sodium sulphate, and with it fill a bottle. Either put in the glass stopple or cover the top with a thin layer of oil, and let the bottle stand. The salt will remain for months without crystallizing; but if it be taken up, and shaken ever so little, the whole mass will instantly form into crystals, so filling the bottle that not a drop of water will escape, even if it be inverted. Should there be any hesitation in crystallizing at the moment, drop into the bottle a minute crystal of the salt, and the effect will instantly be seen in the darting of new crystals in every direction.

† This acid was formerly allowed to escape, causing the destruction of all vegetation in the neighborhood. It is now, however, absorbed so perfectly that the gases which escape from the top of the chimney will not render turbid a solution of silver nitrate (see page 166), showing that there is not a trace of the acid left.

forming Na_2CO_3 and CaS. Out of this mass, called from its color "black-ash," the Na_2CO_3 is dissolved,* and then crystallized, making the "soda-ash" of commerce.

Hydrogen Sodium Carbonate ($HNaCO_3$),† "*Bi-carbonate of Soda*," is the "soda" of the cook-room. It is prepared by the action of CO_2 on sodium carbonate. The CO_2 may be easily liberated by the action of an acid. (See p. 234.)

AMMONIUM.

Symbol, H_4N........Molecular Weight, 18.

THIS is a compound which has never been separated, but it is generally thought to be the base of the salts formed by the action of the acids upon the alkali ammonia, which in form, color, and lustre closely resemble the corresponding salts of K. The analogy between its action and that of the simple metals is so very striking ‡ that it is considered a compound metal, acting the part of a simple one, as Cy does that of a compound halogen (see p. 84). §

* The insoluble residuum of CaS, and the superfluous coal, form around the alkali works a mountain of waste. Attempts have been made to extract the S, and at the Paris exposition large blocks thus obtained were exhibited; but the operation has failed of commercial success.

† The rational formula (see note, page 127) is $Na_2O.H_2O.2CO_2$, whence the name bicarbonate of soda.

‡ When H_3N is dissolved in H_2O, forming $H_3N.H_2O$, the compound may be represented as (H_4N) HO. Comparing this with the formula for caustic potash, KHO, we see that the group of elements H_4N corresponds to the K. Thus we may call a solution of H_3N, ammonium hydrate, as one of potash is a potassium hydrate. Both act as powerful bases, neutralize the acids and form soaps.

§ The following experiment is thought by some to be an additional proof of the metallic nature of this compound substance. Heat moderately in a test-tube half a fluid-dram of Hg with a piece of Na the size of a pea. The two metals

Compounds —*Ammonium Chloride,* H_4NCl,* *Sal-ammoniac,* is prepared from the ammoniacal liquor of the gas-works. (See p. 83.) Its tough fibrous crystals reveal no trace of the pungent ammonia, yet it can easily be set free, as we have already seen (p. 48). Sal-ammoniac is soluble in H_2O. It is used in medicine, in the preparation of H_3N and its salts, in dyeing, and also in soldering, as it dissolves the coating of the oxide of the metal and preserves the surfaces clear for the action of the solder.

Ammonium Carbonate, Sal-volatile, Smelling Salts, is prepared by the action of chalk upon sal-ammoniac.† It is largely used by bakers in raising cake. (See p. 235.)

Ammonium Nitrate ($H_3N,HNO_3=H_4N,NO_3$) may be readily formed by cautiously adding dilute HNO_3 to aqua ammonia until the liquid becomes neutral, and then evaporating. Long, needle-shaped crystals will form. Thus two fiery liquids combine to produce a solid having no resemblance to either of them. By heat this salt may be converted into H_2O and N_2O. (See p. 46.)

will combine, forming a pasty amalgam. When cold, pour over it a solution of sal-ammoniac. The amalgam will immediately swell up to eight or ten times its original bulk, *retaining, however, its metallic lustre.* The ammonium cannot be separated from the amalgam, since, on heating, it decomposes, and on being thrown into water, H is set free and H_3N formed.

* Its rational formula is $H_3N.HCl$, whence it is often called hydrochlorate of ammonia.

† It is a sesquicarbonate, but by the constant loss of H_3N through evaporation, it becomes crusted with a spongy coat of the "bicarbonate," hydrogen ammonium carbonate (H_4N), HCO_3.

METALS OF THE ALKALINE EARTHS.

Ca, Ba, and Sr.

CALCIUM.

Symbol, Ca.... Atomic Weight, 40.... Specific Gravity, 1.57.

Ca exists abundantly in limestone, gypsum, and in the bones of the body.* It commonly occurs as an oxide or a carbonate.

Compounds.—*Calcium Oxide* (CaO), *Caustic* or *Quicklime*, is obtained by heating limestone ($CaCO_3$) in large kilns. The CO_2 is driven off by the heat, and the CaO is left as a white solid.

Fig. 55.

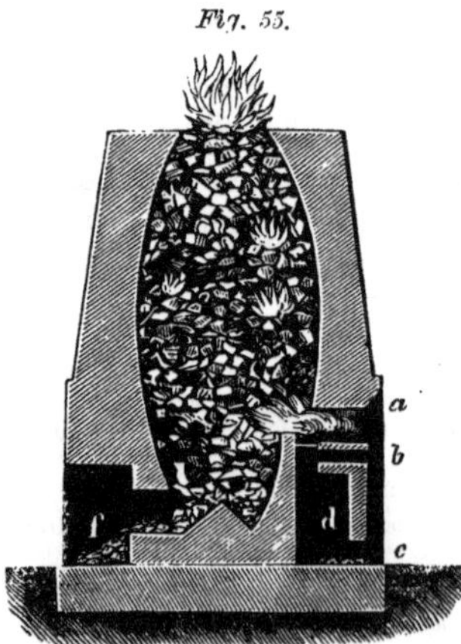

Lime-kiln.

Fig. 55 shows a form of lime-kiln in which the process is continuous. At *a, b, c, d,* are the doors for the fuel, ash-pit, etc. The lime-kiln is fed at the top from time to time, while the lime is taken out at *f* as fast as formed.

Properties.—CaO is a strong alkali, and corrodes the flesh. Its test is CO_2, producing a milky precipitate of

* "There are 5 lbs. of phosphate of lime, one of carbonate of lime, and 3 oz. of fluoride of calcium in the body of an adult weighing 154 lbs."—NICHOLS.

$CaCO_3$. It has such an affinity for H_2O, that fifty-six pounds of lime will absorb eighteen pounds of H_2O, forming CaO,H_2O, or "slacked lime," and expanding to several times its original size, with the evolution of much heat. CaO absorbs H_2O from the air, and then CO_2, thus gradually becoming "air-slacked lime." It is sparingly soluble in water. A thin film of calcium carbonate will soon gather over a solution of lime exposed to the air. *Water-lime* contains a little clay and will harden under water.

Uses.—*Whitewash* is a "milk of lime," *i. e.*, lime diffused through water. *Concrete* is a cement of coarse gravel and water-lime. It is of great durability. *Hard finish* is a kind of plaster in which gypsum is used to make the wall smooth and hard. *Calcimine* is a variety of whitewash made of whiting or plaster of Paris. *Mortar* is a mixture of lime and sand wet with H_2O. It hardens rapidly, by absorbing CO_2 from the air to form a carbonate, and partly, perhaps, by uniting with the SiO_2 of the sand to form a silicate.*

Lime is valuable as a fertilizer. It acts by rapidly decomposing all vegetable matter, and thus forming H_3N for the use of plants.† It also sets free the alkalies

* "If common mortar be protected from the air, it will remain without hardening for many years. It is stated that lime still in the condition of a hydrate has been found in the Pyramids of Egypt. When the ruins of the old castle of Landsberg were removed, a lime-pit, that must have been in existence three hundred years, was found in one of the vaults. The surface was carbonated to the depth of a few inches, but the lime below this was fresh as if just slacked, and was used in laying the foundations of the new building."—*American Cyclopedia.*

† If applied to a compost heap, it will set free H_3N, thus robbing it of its most valuable constituent. This can be saved by sprinkling the pile with dilute H_2SO_4, or plaster, or by mixing it with dry muck, which will absorb the gas. If there is any copperas (produced by the oxidation of iron pyrites) in the soil, the lime will decompose it, forming gypsum and iron-rust, thus changing a noxious ingredient into an element of fertility.

that are combined in the soil, and furnishes them to the plants, becoming itself a carbonate. Lime is also used extensively in the preparation of bleaching powder, in refining sugar, in making candles, in tanning, and in the manufacture of coal-gas.

Fig. 56.

A cave with stalactites and stalagmites.

Calcium Carbonate, $CaCO_3$, includes limestone, chalk, marble, and marl, and forms the principal part of corals, shells, etc. H_2O charged with CO_2 dissolves $CaCO_3$ freely, which, when the gas escapes on exposure to the air, is deposited. In limestone regions, the water trickling down into caverns has formed "stalactites," which depend from the ceiling, and "stalagmites," that rise from the floor. These frequently assume curious and grotesque forms, as in the Mammoth Cave. Around many springs, the water, charged with $CaCO_3$ in solution, flows over moss or some vegetable substance, upon which the stone is deposited. The spongy rock thus formed is called calcareous tufa, or "petrified moss." (See *Geology*, p. 49.) *Marble* is crystallized limestone. *Chalk* or *marl* is a porous kind of limestone, formed from beds of shells, but not compressed as in common limestone. *Whiting* is ground chalk.

Calcium Sulphate ($CaSO_4,2H_2O$), *Gypsum, Plaster, etc.**—This occurs as beautiful fibrous crystals in satin spar, as transparent plates in selenite, and as a snowy-white solid in alabaster. It is soft, and can be cut into rings, vases, etc. When heated it loses its water of crystallization, and is ground into powder, called "Plaster of Paris," from its abundance near that city. Made into a paste with H_2O, it first swells up, and then immediately hardens into a solid mass. This property fits it for use in copying medals and statues, forming moulds, fastening metal tops on glass lamps, etc. *Plaster* (unburned or hydrated gypsum) is used as a fertilizer.† Its action is probably somewhat like that of lime, and in addition it gathers up ammonia and holds it for the plant.

* The rational formula is $CaO.SO_3$; hence it is commonly called "sulphate of lime." Comparing the formula H_2SO_4 and $CaSO_4$, we see that one atom of Ca can replace two atoms of H; it is therefore one of a class of elements called *dyads* (*duo*, two). An atom of K, as we have seen, can displace only one atom of H; it belongs to the *monads* (*monos*, alone).

† It is said that Franklin brought $CaSO_4$ into use by sowing it over a field of grain on the hill-side, so as to form, in gigantic letters, the sentence, "Effects of gypsum." The rapid growth produced soon brought out the words in bold relief, and decided the destiny of gypsum among farmers.

Calcium Sulphite, $CaSO_3$, should be distinguished from the sulphate. It is much used in preserving cider, being sold as "sulphite of lime."

Calcium Phosphate, "*Phosphate of Lime,*" is frequently termed *bone phosphate,* as it is a constituent of bones. (See p. 120.) It is found in New Jersey, South Carolina,* and Canada. It is the valuable part of certain guanos. Fertilizers are prepared by treating ground bones with H_2SO_4, forming the so-called superphosphate of lime.† This is a mixture of gypsum and hydrogen calcium phosphate. The latter furnishes phosphorus to the growing plant to store in its seeds.—*Example:* corn, wheat.

STRONTIUM AND BARIUM.

THESE metals are very like Ca. The salts of Ba give a green tint to a flame and those of Sr a beautiful crimson; and are hence much used in pyrotechny. Barium sulphate, commonly called barytes, is found as a white mineral, noted for its weight, whence it is often termed heavy spar. Indeed, the term barium is derived from a Greek word meaning *heavy.* This mineral is largely used for adulterating white-lead. $BaCl_2$ is a test for H_2SO_4. (See p. 117.)

* Along the coast of South Carolina are millions of tons of rocks holding this important element of plant-food. The phosphatic beds extend over an area of several hundred square miles, and in some cases they are twelve feet thick. It is estimated that from 500 to 1000 tons underlie each acre.—*Fireside Science.*

† Ca_32PO_4 (tricalcium phosphate) + $2H_2SO_4 = H_4Ca2PO_4$ (acid phosphate or superphosphate) + $2CaSO_4$ (calcium sulphate). As the gypsum is only slightly soluble in water, the superphosphate may be removed from the mass by filtering, and used as a fertilizer, or be heated with charcoal to form phosphorus. In that case it is reconverted into tricalcium phosphate while a part of the phosphoric acid breaks up thus: $4H_3PO_4 + 16C = P_4 + 6H_2 + 16CO$.

MAGNESIUM.*

Symbol, Mg....Atomic Weight, 24.3....Specific Gravity, 1.7.

Source.—Mg is found in augite, hornblende, meerschaum, soap-stone, talc, serpentine, dolomite, and other rocks. Its salts give the bitter taste to sea-water. When pure, it has a silvery lustre and appearance. It is very light and flexible. A thin ribbon of the metal will take fire from an ignited match, when it will burn with a brilliant white light, casting dense shadows through an ordinary flame, and depositing flakes of MgO. This light possesses the actinic or chemical principle so perfectly, that it is used for taking photographs at night, views of coal mines, interiors of dark churches, etc. It has every ray of the spectrum, and so does not, like gas-light, change some of the colors of an object upon which it falls. Magnesium lanterns are much used for purposes of illumination. By means of clockwork, the metal, in the form of a narrow ribbon, is fed in front of a concave mirror, at the focus of which it burns. It is hoped that the process of manufacture † may be cheapened, so that Mg may be furnished at a rate which will bring it within the scope of the arts.

Compounds. — *Magnesium Carbonate,* $MgCO_3$,

* Mg is now usually classified with Zn, Cd, and In, since, while the metals of the alkaline earths decompose H_2O with avidity and set H free, these four act only upon steam at a red heat, being without effect upon H_2O at ordinary temperatures. Mg is treated here for convenience, while Zn is described among the common or useful metals. Cd and In are of no practical value. The oxide of Mg has a slight alkaline reaction, and until recently, Mg was considered one of the metals of the alkaline earths.

† It is now prepared by heating $MgCl_2$ with metallic Na.

is the "magnesia alba" or common magnesia of the druggist. *Magnesium sulphate* ($MgSO_4,7H_2O$) is known as Epsom salt, from a celebrated spring in England in which it abounds.

ATOMIC EQUIVALENCE.*

The Quantivalence of an Element is its power of combining with H. Thus, if we examine the compounds HCl, H_2O, H_3N and H_4C, we find the molecule of each has respectively 1, 2, 3, 4 atoms of H; and we conclude that the elements Cl, O, N and C have different powers of combination whereby they can unite with a different number of H atoms. The same principle holds throughout all the elements. They are therefore divided into classes according to their power of combining with H, and are termed respectively, monatomic, diatomic, triatomic, &c., or monads, dyads, triads, &c. (Note, p. 139.) Each atom is supposed to have a certain number of bonds or more probably poles, like those of a magnet, by which it holds or becomes united to other atoms. H, Na, Cl have but one bond or pole and can therefore, like all monads, only form pairs. O has two poles and can combine with H_2, an atom of H at each pole, or with K_2 or HCl, or with an atom of some other dyad, as Zn. (Table, p. 288.) C has four poles and can unite with H_4 or Cl_4, or O_2, or Cl_2O or H_2O; *i. e.* with four atoms of a monad, two of a dyad, or one of a dyad and two of a monad. The quantivalence of an element is commonly represented thus, Cl^{I}, O^{II}.

* For a full explanation of this subject see that delightful work Cooke's New Chemistry, Hoffman's Modern Chemistry or Wurtz's Introduction to Chemistry.

METALS OF THE EARTHS.

Al, G, E, Y, Ce, La, D.

ALUMINUM.

Symbol, Al....Atomic Weight, 27.5....Specific Gravity, 2.6.

Source.—Al is named from alum, in which it occurs. It is also called the "clay metal." It is the metallic base of clay, mica, slate, and feldspar rocks. Next to O and Si, it is probably the most abundant element of the earth's crust. It is a bright, silver-white metal; does not oxidize in the air, nor tarnish by H_2S. It gives a clear musical ring; is only one-fourth as heavy as Ag; is ductile, malleable, and tenacious. It readily dissolves in HCl, and in solutions of the alkalies, but with difficulty in HNO_3 and H_2SO_4. On account of its abundance (every clay-bank is a mine of it) and useful properties, it must ultimately come into common use in the arts and domestic life.

Compounds.—*Aluminum Oxide* (Al_2O_3).—*Alumina*, crystallized in nature, forms valuable Oriental gems. They are variously colored by the oxides;—blue, in the sapphire; green, in the emerald; yellow, in the topaz; red, in the ruby. Massive, impure alumina combined with magnetic iron, is called *emery*, and used for polishing.

Aluminum Silicate ($Al_2O_3,2SiO_2$), *Silicate of Alumina, Common Clay.*—When the clay rocks decay, by the resistless and constant action of the air, rain, and frost, they crumble into soil. This contains clay, silica, and other impurities, such as lime, magnesia, oxide of iron, etc. The clay gives firmness to the soil, and retains moisture, but is cold and tardy in producing vegetable growth. When free from Fe, it is used for making tobacco-pipes. When colored by ferric oxide, it is known as ochre, and is employed in painting. Common stone and red earthen-ware are made from coarse varieties of clay; porcelain and china-ware require the purest material. Fire-bricks and crucibles are made from a clay which contains much SiO_2. Fullers' earth is a very porous kind, and by capillary attraction absorbs grease and oil from cloth.

Glazing.—When any article of earthen-ware has been moulded from clay, it is baked. As the ware is porous, and will not hold H_2O, a mixture of the coarse materials from which glass is made is then spread over the vessel, and heated till it melts and forms a glazing upon the clay. Ordinary stone-ware is glazed by simply throwing damp NaCl into the furnace. This volatilizes, and being decomposed by the hot clay makes a sodium silicate over the surface, while fumes of HCl escape. Pb is sometimes used to give a yellowish glaze, which is very injurious, as it will dissolve in vinegar, and form sugar of lead, a deadly poison. The color of pottery-ware and brick is due to the oxide of iron present in the clay. Some varieties have no iron, and so form white ware and brick.

Fig. 58.

Baking Porcelain.

Alum is made by treating clay with H_2SO_4, forming

an aluminum sulphate. On adding potassium sulphate a double salt is produced, which separates in beautiful octahedral crystals ($Al_2K_24SO_4 + 24H_2O$). Instead of K an ammonium salt* is now generally added, and an ammonium alum made, which takes the place of the former in the market.† Alum is much used in dyeing. It unites with the coloring matter, and binds it to the fibres of the cloth. It is therefore called a mordant (*mordeo*, I bite).

SPECTRUM ANALYSIS.

MANY of the metals named as rare have been recently discovered by what is termed Spectrum Analysis. We have already noticed that various metals impart a peculiar color to flame; thus Na gives a yellow tinge, copper a green, etc. If now we look at these colored flames through a prism, we shall find, instead of the "spectrum" we are familiar with, a dark space strangely ornamented with bright-tinted lines. Thus the spectrum of Na has one double, yellow line;‡ Ag, two green lines; Cs, a beautiful blue line. Each metal makes a distinctive spectrum, even when the flame is colored by several substances at once. This method of analysis is so deli-

* Ammonium sulphate, from the ammoniacal liquor of the gas-works. (See page 83.)

† There are a large number of other alums known, in which the isomorphous sesquioxides of iron, chromium, and manganese are substituted for the alumina in common alum: all these alums occur in regular octahedra, and cannot be separated by crystallization when present in solution together.

‡ The yellow, sodium line consists of two lines lying so closely together as to seem as one. They correspond to Fraunhofer's lines D (see Frontispiece, No. 2), as given in the drawings of Kirchhoff and Bunsen.

cate that $\frac{1}{180,000,000}$ of a grain of Na, or $\frac{1}{6,000,000}$ of L, can be detected in the flame of an alcohol lamp;* while a substance exposed to the air for a moment even will give the Na lines from the dust it gathers. L has thus been found to exist in tea, tobacco, milk, and blood, although in such minute quantities as to have eluded detection by former methods of analysis.

PRACTICAL QUESTIONS.

1. In the experiment with Na_2SO_4 on page 133, an accurate thermometer will show that in making the solution, the temperature of the liquid will fall, and in its solidification, will rise. Explain.

2. If, in making the solution of Na_2SO_4, we use the salt which has effloresced, and so become anhydrous, the temperature will rise instead of falling as before. Explain.

3. Why is KNO_3 used instead of $NaNO_3$ for making gunpowder?

4. Why is a potassium salt preferable to a sodium one in glass-making?

5. What is the glassy slag so plentiful about a furnace?

6. State the formulæ of nitre, saleratus, carbonate and bicarbonate of soda, plaster, pearlash, saltpetre, plaster of Paris, gypsum, carbonate and bicarbonate of potash, sal-soda, and soda.

7. Explain how ammonium carbonate is formed in the process of making coal-gas.

8. Upon what fact depends the formation of stalactites?

9. Why is HF kept in gutta-percha bottles?

10. Explain the use of borax in softening hard water?

11. How are petrifactions formed?

12. In what part of the body, and in what forms, is phosphorus found?

13. Why are matches poisonous? What is the antidote? (See *Physiology*, page 209.)

14. Will the burning phosphorus ignite the wood of the match?

* For the more perfect examination of the spectra, a "spectroscope" is used. This consists of a tube with a narrow slit at one end, which lets only a single ray of colored light fall upon the prism within, and at the other a small telescope, through which one can look in upon the prism and examine the spectrum of any flame. (See *Astronomy*, page 285.)

15. What philosophical principle is illustrated in the ignition of a match by friction?

16. How much H_2O would be required to dissolve a pound of KNO_3?

17. What causes the bad odor after the discharge of a gun?

18. Write in parallel columns (see *Question* 54, page 96) the properties of common and of red phosphorus.

19. What causes the difference between fine and coarse salt?

20. Why do the figures in a glass paper-weight look larger when seen from the top than from the bottom?

21. What is the difference between water-slacked and air-slacked lime?

22. Why do oyster-shells on the grate of a coal-stove prevent the formation of clinkers?

23. How is lime-water made from oyster-shells?

24. Why do newly-plastered walls remain damp so long?

25. Will lime lose its beneficial effect upon a soil after frequent applications?

26. What causes plaster of Paris to harden again after being moistened?

27. What is the difference between sulphate and sulphite of lime?

28. What two classes of rays are contained in the magnesium light?

29. What rare metals would become useful in the arts, if the process of manufacture were cheapened?

30. What is the rational formula for calcium carbonate? Calcium sulphite? Calcium sulphate?

31. Why is lime placed in the bottom of a leach-tub?

32. Is saleratus a salt of K or of Na?

33. Why will Na burst into a blaze when thrown on hot water?

34. Why are certain kinds of brick white?

35. Illustrate the force of chemical affinity.

THE USEFUL METALS.

IRON.

Symbol, Fe....Atomic Weight, 56....Specific Gravity, 7.8.

IRON is the symbol of civilization. Its value in the arts can be measured only by the progress of the present age. In its adaptations and employments it has kept pace with scientific discoveries and improvements, so that the uses of iron may readily indicate the advancement of a nation. It is worth more to the world than all the other metals combined. We could dispense with gold and silver—they largely minister to luxury and refinement, but iron represents solely the honest industry of labor. Its use is universal,* and it is fitted alike for massive iron cables, and for screws so tiny that they can be seen only by the microscope, appearing to the naked eye like grains of black sand.

Its abundance everywhere indicates how indispensable the Creator deemed it to the education and development

* "Iron vessels cross the ocean,
Iron engines give them motion,
Iron needles northward veering,
Iron tillers vessels steering,
Iron pipe our gas delivers,
Iron bridges span our rivers,
Iron pens are used for writing,
Iron ink our thoughts inditing,
Iron stoves for cooking victuals,
Iron ovens, pots, and kettles,
Iron horses draw our loads,
Iron rails compose our roads,
Iron anchors hold in sands,
Iron bolts and rods and bands,
Iron houses, iron walls,
Iron cannon, iron balls,
Iron axes, knives, and chains,
Iron augers, saws, and planes,
Iron globules in our blood,
Iron particles in food,
Iron lightning-rods on spires,
Iron telegraphic wires,
Iron hammers, nails, and screws,
Iron everything we use."

of man. There is no "California" of iron. Each nation has its own supply. No other material is so enhanced in value by labor.

1 lb.	good iron is worth, say..............	$ .04
1 "	bar steel..........................	.17
1 "	inch-screws........................	1.00
1 "	steel wire.........................	3 to 7.00
1 "	sewing-needles.....................	14.00
1 "	fish-hooks.........................	20 to 50.00
1 "	jewel-screws for watches............	3,500.00
1 "	hair-springs for American watches...	16,000.00 *

Source.—Fe is rarely found native, *i. e.*, in the metallic condition. Meteors, however, containing as high as 93 per cent. of Fe associated with Ni and other metals, have fallen to the earth from space. Fe in combination with various other substances is widely diffused. It is found in the ashes of plants and the blood † of animals. Many minerals contain it in considerable quantities. The ores from which it is extracted are generally oxides or carbonates.

Preparation.—*Smelting of Iron Ores.*—Fe is locked up with O in an apparently useless stone. C is the key that is ready made and left for our use by the Creator. The process adopted at the mines is very simple. A tall blast-furnace is constructed of stone and lined with fire-brick. At the top is the door, and at the bottom are pipes for forcing in hot air, sometimes twelve

* One pound (Troy) of fine gold is worth in standard coin $248.062. All the above statements are based on careful and actual valuation.

† There are only about 100 grains of Fe in the blood of a full-grown person—about enough to make a ten-penny nail—yet it gives energy and life to the system. The metal is often administered as a tonic in the form of a fine powder, or a citrate of iron, and is a powerful remedy.

thousand cubic feet per minute, by means of pistons driven by steam-power. The furnace, being filled with

Fig. 59.

A Blast-Furnace.

limestone, coal and iron ore, in alternate layers, the fire is ignited. The C* unites with the O of the ore, and

* A little N sometimes unites with some C and K, forming potassium cyanide, or with Ti, if any is present, making beautiful copper-colored crystals of titanium cyanide.

goes off as CO_2. The $CaCO_3$ forms with the SiO_2 and other impurities a richly-colored glassy slag, which rises to the top. The melted Fe runs to the bottom, and is drawn off in channels cut in the sand on the floor of the furnace. The large main one is called the *sow*, and the smaller lateral ones the *pigs*, and hence the term *pig-iron*.

Fig. 60.

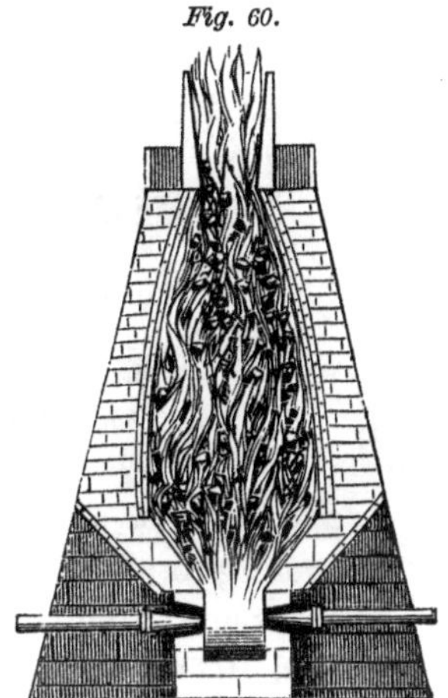

Section of a Blast-Furnace.

Varieties of Fe.—The usual forms are *cast*, *wrought*, and *steel*, depending upon the proportion of C which they contain. Cast-iron has from 2 to 5 per cent., steel from 1 to 2 per cent., and wrought-iron about $\frac{1}{5}$ per cent.

1. ***Cast*** Fe is the form which comes from the furnace. It is brittle, cannot be welded, and is neither malleable nor ductile. It is an exception to the law that "cold contracts," since at the instant of solidification it expands, so as to copy exactly every line of the mould into which it is poured. This fits it perfectly for castings. These may be made so soft as to be easily turned and filed, or so hard, by cooling in iron moulds,* that no tool will affect them.

2. ***Wrought or Malleable*** Fe is made by burning the C from cast-iron, in a current of highly-heated air, in what is called a reverberatory furnace. The Fe is stirred

* These moulds are called "chills," and the iron is termed chilled iron. It is used for burglar-proof safes.

constantly, and exposed to the heated air by means of long "puddling-sticks," as they are termed. It is taken out while white-hot and beaten under a trip-hammer to force out the slag; and lastly, pressed between grooved rollers to bring the particles of Fe nearer each other and give it a fibrous structure.* It is now malleable and ductile, and can be welded.† Fe is hardened by cooling rapidly, and softened by cooling slowly. The blacksmith tempers his work by plunging the article in cold H_2O.

Fig. 61.

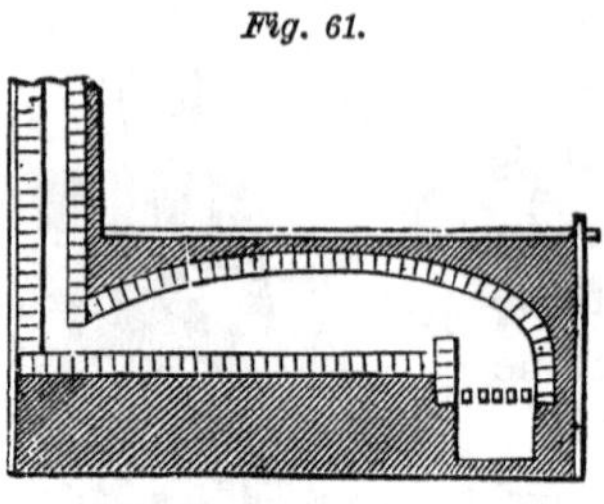

A Reverberatory Furnace.

3. ***Steel*** contains less C than cast, and more than wrought, iron. It is therefore made from the former by burning out a part of the C, and from the latter by heating in boxes of charcoal, and so adding C.‡ The value of steel depends largely upon its *temper*. This is determined by heating the article and then allowing it to cool. The higher the temperature the softer the steel. The

* This fibrous structure is so noticeable that if a bar of the best Fe be notched with a chisel and then broken by a steady pressure, the fracture will present a stringy appearance, like that of a green stick. By constant jarring, however, Fe tends to take a crystalline structure, becoming rotten and brittle, so that cannon, the axles of cars, etc, are condemned after a certain time, although no flaw may appear.

† It has been beaten into leaves so thin that they have been used for writing-paper—six hundred leaves being only half an inch in thickness—and has been drawn into wire as fine as a hair.

‡ Cheap knives made of soft iron are often covered with a superficial coating of steel in this way. When we use such knives, we soon wear through this crust, and find metal beneath which will take no edge. This is termed *case-hardening*.

workman decides this by watching the color of the oxide which forms on the surface.* Razors require a straw yellow; table-knives, a purple; springs and swords, a bright blue; and saws, a dark blue tint.†

Bessemer's Process is now extensively used for making steel. Several tons of the best pig-iron are melted, and poured into a large crucible hung on pivots so as to be easily tilted. Hot air driven in from beneath, bubbles up through the liquid mass, producing an intense combustion. The roar of the blast, the hot, white flakes of slag ever and anon whirled upward, the long flame streaming out at the top, variegated by tints of different metals, and full of sparks of scintillating iron, all show the play of tremendous chemical forces. The operation lasts about twenty minutes, when the Fe is purified of its C and Si. Enough spiegel-eisen (looking-glass iron), an ore rich in C and Mn, is added to convert it into steel, when it is poured out and cast into ingots.‡

* The thin pellicles of iron-rust on standing H_2O produce a beautiful iridescent appearance in the same way, the color changing with the thickness of the oxide. Just so a soap-bubble exhibits a play of variegated colors according to the thickness of the film in different parts. (See "Interference of Light," *Physics*, p. 168.)

† These colors are removed in the subsequent processes of grinding and polishing, but they may be seen in a handful of old watch-springs, to be obtained of any jeweller.

‡ In 1760, there lived at Attercliffe, near Sheffield, a watchmaker named Huntsman. He became dissatisfied with the watch-springs in use, and set himself to the task of making them homogeneous. "If," thought he, "I can melt a piece of steel and cast it into an ingot, its composition should be the same throughout." He succeeded. His steel became famous, and Huntsman's ingots were in universal demand. He did not call them cast-steel. That was his secret. The process was wrapped in mystery by every means. The most faithful men were hired. The work was divided, large wages paid, and stringent oaths taken. One midwinter night, as the tall chimneys of the Attercliffe steelworks belched forth their smoke, a belated traveler knocked at the gate. It was bitter cold; the snow fell fast; and the wind howled across the moor. The stranger, apparently a common farm-laborer seeking shelter from the storm,

Compounds.—1. *Black or Magnetic Oxide* (Fe_3O_4) is found in the loadstone, Swedish iron-ore, scales which fly off in forging iron, and in mines in various parts of the United States. It is the richest of the ores and contains as high as 72 per cent. of the metal. 2. *Red Oxide of Iron*, sesquioxide (ferric oxide, Fe_2O_3), is seen in red iron-ore, in the beautiful radiated and fibrous specimens of hematite,* specular † iron, red ochre and chalk, bricks and pottery-ware. The sesquioxide, combining with H_2O, forms—3. *Hydrated Sesquioxide of Iron* (ferric hydrate, $Fe_2O_3,3H_2O$). This has a brown or yellow color, which changes to red by heat when the water is expelled, as in the burning of brick, pottery-ware,‡ etc. These oxides generally give the brown, yellow, or red tints seen in sand, gravel, etc. The ferric oxide and hydrate are remarkable for the facility with which they absorb O from the air, and impart it to other bodies. This is familiar in the rusting of nails in clap-boards, hinges in gate-posts, hooks in ropes, etc, etc.

Iron Carbonate, $FeCO_3$, is found as spathic § and

awakened no suspicion. The foreman, scanning him closely, at last granted his request and let him in. Feigning to be worn-out with cold and fatigue, the poor fellow sank upon the floor and was soon seemingly fast asleep. That, however, was far from his intention. Through cautiously opened eyes, he caught glimpses of the mysterious process. He saw workmen cut bars of steel into bits, place them in crucibles, which were then thrust into the furnaces. The fires were urged to their utmost intensity until the steel melted. The workmen, clothed in rags, wet to protect them from the tremendous heat, drew forth the glowing crucibles and poured their contents into moulds. Huntsman's factory had nothing more to disclose. The secret of cast-steel was stolen.

* *Hæmatites*, blood-like, from the red color of its powder.

† *Speculum*, a mirror, from the brilliant lustre of its steel-gray crystals and mica-like scales in micaceous iron-ore.

‡ Clay, containing ferrous oxide (FeO), becomes red by its conversion into ferric oxide.

§ *Spath*, spar, as some specimens consist of transparent, shiny crystals, having the same form as calcareous spar (calcium carbonate).

clay ironstone, and often contains some manganese,* which fits it for the manufacture of certain kinds of steel, whence it is termed steel-ore. In chalybeate springs, the free CO_2 in the water holds the $FeCO_3$ in solution. On coming to the air, the CO_2 escapes, and the Fe, absorbing O, is deposited as hydrated ferric oxide, forming the ochry deposit so common around such springs.

Iron Disulphide (FeS_2), *Iron Pyrites, Fool's Gold*—so called, because it is often mistaken by ignorant persons for Au. It occurs in cubical crystals and bright shiny scales. It can be easily tested by roasting on a hot shovel, when we shall catch the well-known odor of the SO_2. FeS_2 is used as a source of S, and also in the manufacture of H_2SO_4.

Ferrous Sulphate ($FeSO_4,7H_2O$), *Green Vitriol, Copperas,* is made by the action of H_2SO_4 on Fe, and, at Stafford, Connecticut, from FeS_2, by exposure to air and moisture. It is used in dyeing, making ink, and in photography.

* Manganese is a hard, brittle metal, resembling cast-iron in its color and texture. It takes a beautiful polish. Its binoxide, the black oxide of manganese, is used in the manufacture of O, Cl, etc. By fusing MnO_2, $KClO_3$, and KHO, a dark, green mass is obtained called "*chameleon mineral.*" It contains potassium manganate. If a piece of this be placed in H_2O, the solution will undergo a beautiful change from green, through various shades, to purple. This is owing to the gradual formation of permanganic acid. The change may be produced instantaneously by a drop of H_2SO_4. Potassium permanganate is remarkable for the facility with which it parts with its O, and thereby loses its color. It is used extensively as a disinfectant, and as a test of the presence of organic matter. (See page 60.)

ZINC.

Symbol, Zn....Atomic Weight, 65....Specific Gravity, 7.15.
Fusing Point, 773° F.

Source.—Zn, or "spelter," as it is called in commerce, is found as ZnO, or red oxide, in New Jersey, and as ZnS, or zinc blende, in many places.

Fig. 62.

Roasting Zinc Ore.

Preparation.—ZnO is smelted on the same principle as iron ore, by heating with C. The reaction is as follows: $ZnO + C = Zn + CO$. Both these products distil, the Zn vapor being condensed while the CO gas escapes.

Properties.—Zn is ordinarily brittle, but when heated to 200° or 300° F., it becomes malleable, and can be rolled out into the sheet Zn in common use. It burns in the air with a magnificent green light, forming flakes of ZnO, sometimes called "Philosopher's Wool."* When exposed to the air Zn soon oxidizes, and the thin film of white oxide formed over the surface protects it from further change.

Uses.—Its economic uses are familiar. Sheet Fe dipped in melted Zn forms what is termed *galvanized iron.* Water-pipes made of this material are as unsafe as lead (see p. 160) until the Zn is entirely corroded.

* *Example:* On a red-hot ladle, sprinkle some powdered saltpetre and Zn filings. The KNO_3 will furnish O, and the metal will burn with great brilliancy.

The oxide and carbonate of zinc are rapidly formed, and these poisonous salts remain in the H_2O. There is the same objection to metallic-lined ice-pitchers. Galvanic action between the metals promotes corrosion. H_2O standing in reservoirs lined with Zn should not be used for drinking purposes. In the case of zinc-covered roofs the rain-water contains zinc oxide.*

Compounds.—*Zinc Oxide,* ZnO, is sold as zinc-white, and is valued as a paint, since it does not blacken by H_2S like white-lead; but it is quite as hurtful to the painter. *Zinc sulphate* ($ZnSO_4$), *white vitriol,* is used in medicine.

TIN.

Symbol, Sn.... Atomic Weight, 118.... Specific Gravity, 7.2.
Fusing Point, 442° F.

Source.—Sn, though one of the metals longest known to man, is found in but few localities. It is reduced from its binoxide by the action of C.

Properties.—It is soft and not very ductile, but is quite malleable, so that tinfoil is not more than $\frac{1}{1000}$ of an inch in thickness. When quickly bent, it utters a shrill sound, called the "tin cry," caused by the crystals moving upon each other. Sn does not oxidize at ordinary temperatures. Its tendency to crystallize is remarkable.†

* When they were first introduced in Boston the washerwomen complained that the rain-water was hard, decomposed the soap, and made their hands crack.

† *Example:* Heat a piece of Sn till the coating begins to melt; then cool quickly in H_2O and clean in dilute aqua-regia. The surface will be found covered with beautiful crystals of the metal.

Uses.—Common sheet-tin is formed by dipping sheet-iron in melted Sn, which produces an artificial coating of the latter metal. If we leave H_2O in a tin dish, the yellow spots soon betray the presence of Fe. Pins made of brass wire are boiled with granulated tin, cream of tartar, and H_2O, which give a bright white surface to the metal.*

COPPER.

Symbol, Cu....Atomic Weight, 63.5....Specific Gravity, 8.9.
Fusing Point, 1994° F.

Source.—Cu is found native near Lake Superior, frequently in masses of great size. In these mines stone hammers have been discovered, the tools of a people older than the Indians, who probably occupied this continent, and worked the mines. In the western mounds, also, copper instruments are found. The sulphide, copper pyrites, is a well-known ore. *Malachite* ($CuCO_3,CuO,H_2O$), the green carbonate, admits of a high polish, and is made into ornaments of exquisite beauty.

Properties.—Cu is ductile, malleable, and an excellent conductor of heat and electricity. Its vapor gives a characteristic and beautiful green color to flame. It is hardened by hammering, and softened by heating and plunging into cold H_2O.† HNO_3 is the solvent of Cu. Its test

* The pins are stuck in papers, as we see them, by machinery which picks them up out of a miscellaneous pile, counts them, and inserts them in the paper, ready for the market. The first part of the process is performed by a sort of coarse comb, which is thrust into the heap, and gathers up a pin in each of the spaces between the teeth.

† The reverse of Fe, which fact ruins any theory we might form as to the cause in either case.

is H_3N, forming in a solution an azure-blue precipitate, which dissolves in an excess of the reagent.

Compounds.—*Copper Acetate, Verdigris,** is produced when we soak pickles in brass or copper kettles; the green color which results is caused by this salt—a deadly poison. Preserved fruits, etc., should never stand in such vessels, as the vegetable acids dissolve Cu readily.

Copper Oxide, CuO, is the black coating which collects on copper or brass kettles, and is very poisonous. It dissolves readily in fats and oils. Such utensils should therefore be used only when perfectly bright, and never with fruits, sweetmeats, jellies, pickles, etc.

Copper Sulphate ($CuSO_4,5H_2O$), *Blue Vitriol,* is much used in dyeing, calico printing, and galvanic batteries.

LEAD.

Symbol, Pb. . . . Atomic Weight, 207. . . . Specific Gravity, 11.36.
Fusing Point, 620° F.

Source.—The most common ore of Pb is galena, PbS, which is reduced by roasting in a reverberatory furnace. The S burns and leaves the metal.

Properties.—Pb is malleable, but contracts as it solidifies; so it cannot be used for castings. It is poisonous, though not immediately, as "bullets have been swallowed, and then thrown off without any harm except the fright." Its effects seem to accumulate in the system, and finally

* The term verdigris is sometimes incorrectly applied to the green coating of carbonate, which gathers upon brass or copper in a damp atmosphere.

to manifest themselves in some disease. Persons who work in lead, as painters and plumbers, after a time suffer with colics, paralysis, etc.

Uses.—Pb is much used for water-pipes, and is the most convenient of any metal for that purpose. Pure H_2O passing through the pipe will not corrode the Pb, but the O of the air it contains forms an oxide of lead which dissolves in the H_2O, leaving a fresh surface for oxidation. If there are any sulphates or carbonates in the H_2O, they will form a coating over the Pb, and protect it from further corrosion; and as carbonate of lime is common in hard water, that is generally safe. If, when we examine a lead pipe that is in constant use, we find it covered with a white film, it is a good sign; but if it is bright, there is cause for alarm. Still, however much may be said upon the danger, people will use lead pipes, and the following precautions should be observed: *Always let the water run long enough in the morning before using, to remove all which has remained in the water-pipes during the night;* and when the H_2O is let on again after it has been shut off for a while, *leave the faucet open until the pipe is thoroughly washed.*

The Test of Pb is H_2S, forming lead sulphide, PbS. The following is an interesting illustration: Thicken a solution of lead acetate with a little gum-arabic, so as not to flow too readily from the pen, and then make any sketch which your fancy may suggest. This, when dry, will be invisible. When it is to be used, dampen the paper slightly on the wrong side, and then direct against it a jet of H_2S. The picture will at once blacken into distinctness.

Compounds.—*Lead Oxide,* PbO, the well-known

litharge, is formed by heating Pb in a current of air.* It is used in glass-making, in paints, and in glazing earthenware.

Lead Dioxide, PbO_2, is formed by oxidizing PbO. A mixture of the two, called *minium* or *red-lead,* is used for coloring sealing-wax red, and as a paint.

Lead Carbonate, ($PbCO_3$), *White-Lead.*—This salt is made in large quantities in the following manner: Thousands of earthen pots fitted with covers and containing weak vinegar (acetic acid) and a small roll of Pb, are arranged in immense piles, and then covered with tan-bark. The acetic acid combines with the Pb, but the CO_2 formed by the decomposing tan-bark creeps in under the cover, driving off the acetic acid, and forming lead carbonate. The acetic acid, thus dispossessed, attacks another portion of the Pb, but is robbed again; and so the process goes on, until at last the Pb is exhausted. White-lead is largely adulterated with heavy spar, gypsum, etc.

Fig. 63.

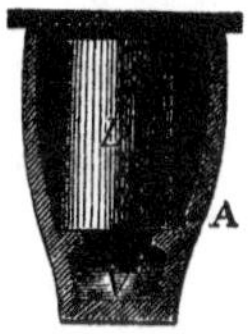

A.—*An earthen pot.*
L.—*A coil of lead.*
V.—*A solution of vinegar.*

Fig. 64.

The Lead-tree.

Lead Acetate, *Sugar of Lead,* has a sweet, pleasant taste, but is a virulent poison. Its antidote is Epsom salt, which forms an insoluble lead sulphate. H_2O dissolves sugar of lead readily. If a piece of Zn, cut in small strips, be suspended in a bottle filled with a solution of lead acetate, the Pb will be deposited upon it by voltaic action in beautiful metallic spangles, forming the "lead-tree."

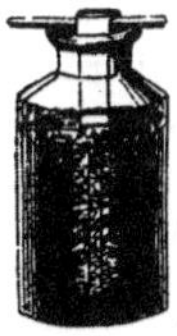

* *Example:* Heat a bit of lead upon charcoal in the oxidizing flame of the blow-pipe. A film of the suboxide forms first, then a yellow crust of the protoxide.

THE NOBLE METALS.

Au, Ag, Pt, Hg, Pd, Ir, Os, Ru, and Ro.

GOLD.

Symbol, Au....Atomic Weight, 197....Specific Gravity, 19.34.
Fusing Point, about 2015° F.

Sources.—Au is found sometimes in masses called nuggets, but generally in scattered grains, or scales. As the rocks in which it occurs disintegrate by the action of the elements and form soil, the Au is gradually washed into the valleys below, and thence into the streams and rivers, where, owing to its specific gravity, it settles and collects in the mud and gravel of their beds.*

Preparation.—As the metal is thus found native, the process is purely mechanical, and consists simply in washing out the dirt and gravel in wash-pans, rockers, sluices,† etc., at the bottom of which the Au accumulates. In the quartz-mills, the rock is thrown into troughs of water, where, by heavy stamps, the ore is crushed to powder.

* In California, Au is found in the detritus (small particles of rock worn off by attrition) of granite and quartz. It occurs in the gravel of hills from the surface to the "bed-rock," sometimes a depth of 300 to 500 feet; in the alluvial soil of the plains, and even in vegetable loam among the roots of grass.

† Sluices are generally used in California. These are gently inclined troughs, sometimes extending for miles. Across the bottom are fastened low wooden bars, called *riffles*, above which quicksilver is placed. The dirt is shovelled into these sluices, or the auriferous hills are cut down, dissolved, and washed through them by powerful streams of water, which are constantly running. The H_2O floats off the debris, while the Hg catches the gold.

As the thin liquid mud thus formed splashes up on either side, it runs over broad, metallic tables covered with Hg; or is washed through a fine wire-screen, and carried to the "amalgamating-pans" by a little stream of water. The Hg unites with the particles of Au and forms with them an *amalgam* (a compound of mercury and a metal). Au is easily separated from Hg by distillation,* and the latter collected to be used again.

Quartation.—Au is commonly found alloyed with Ag. The Ag is then dissolved out by HNO_3. There must be at least three parts of Ag to one of Au, else the gold will protect the silver from the action of the acid. If there is not so much, some is fused with the alloy.†

Properties.—Pure Au is nearly as soft as Pb. It is extremely malleable ‡ and ductile. Its solvent is aqua-regia. It does not oxidize at any temperature, and on account of its indestructibility, it was anciently called the king of the metals.

* The larger part of the Hg is separated from the amalgam by pressure in canvas or buckskin bags, the liquid Hg escaping through the pores, while the amalgam is left quite dry. The latter is then "retorted" for distillation.

† "In works for the refining of gold and silver, the processes can be conducted economically only when great care is taken to avoid the loss of any particles of the precious metals. Thus all the old crucibles are ground and treated with mercury, and after as much gold and silver as possible have been extracted, the residues are sold to the *sweep-washers*, who extract a little more by melting with lead. The very dust off the floors is collected and treated in a similar way."—BLOXAM.

‡ For a description of the process of making gold-leaf, see *Physics*, page 20. "When one of these leaves is held up to the light, it exhibits a beautiful green color, and if it be rendered still thinner, either by beating, or by floating it upon a very weak solution of potassium cyanide, which slowly dissolves it, it transmits, when taken upon a glass plate and held up to the light, a blue, violet, or red light, in proportion as its thickness diminishes. Even when it is so transparent that one may read through it, the yellow color and lustre of the gold are still visible by reflected light. These varying colors of finely-divided gold are turned to account in the coloring of glass and in painting on porcelain."

SILVER.

Symbol, Ag.... Atomic Weight, 108.... Specific Gravity, 10.5.
Fusing Point, 1873° F.

Sources.—Silver is found throughout the West in a great variety of forms—most commonly, however, combined with S, as *black sulphide*, Ag_2S; with Cl, forming

Fig. 65.

Separation of Pb *from* Ag. (*See Bloxam's Metals.*)

horn-silver, AgCl; with S and As or Sb, making *ruby-silver*, and also associated with Pb in ordinary galena.

Preparation.—1st. The *sulphide* is refined as follows: The ore is crushed into fine powder and then roasted with common salt. The Cl of the salt unites with the Ag, forming silver chloride. This is next put into a revolving cylinder with H_2O, Hg, and iron scraps. The Fe removes the Cl from the silver, when the Hg takes it up, thus forming an amalgam of Hg and Ag. From this the Ag is easily obtained, as in gold-washing.* 2d. From *horn-silver*, AgCl, the process is like the latter part of that just described. 3d. From *lead* the Ag can be profitably obtained when there are only two or three ounces in a ton. The alloy of the two metals is melted and then slowly cooled. Pb solidifies much sooner than Ag, and by skimming out the crystals of Pb as fast as formed, it may be almost entirely separated. (See Fig. 65.)

Fig. 66.

A Cupel.

Cupellation.—A cupel (*cupella*, a small cup) is a shallow vessel, made of bone-ashes. In this the Ag, debased with Pb and other impurities, is exposed to a red heat, so as to melt the metals, while a current of hot air plays upon the surface. The Pb oxidizes to PbO, and is absorbed by the porous cupel. The mass appears soiled and tarnished, but the refiner keeps his eye upon it as the process continues, watching eagerly, until at last there is a brilliant play of colors—he catches his own image in the perfect metallic mirror, and the little "button" of pure silver lies gleam-

* The process of reducing silver ores at the West is unlike the German method given above, and varies in different localities. One plan is as follows: The powdered and roasted Ag_2S is placed with Hg in iron pans, five feet in diameter and two feet deep. Here it is kept heated by steam to 180° and agitated by revolving stirrers. The chloride is not roasted, but is simply powdered, and then worked in the pans for an hour with NaCl before adding the Hg.—STEVENSON.

ing at the bottom.* This must now be immediately removed, or it will oxidize and waste.†

Fig. 67.

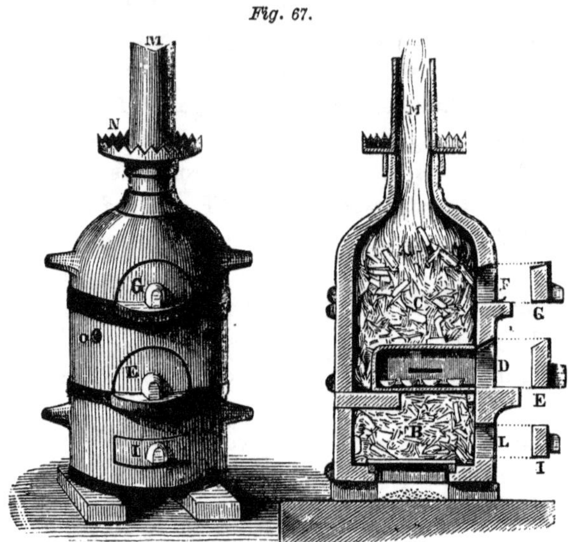

Cupels in Furnace.

Properties.—Ag is the whitest of the metals. It is malleable and ductile. It expands at the moment of solidification, and, therefore, can be cast. It has a power-

* See Malachi iii. 3.

† During the cooling of the cake of Ag, some very remarkable phenomena are observed. When a thin crust of metal has formed upon the surface, the Ag beneath it assumes the appearance of boiling, and the crust is forced up into hollow cones about an inch high, through which the melted Ag is thrown out with explosive violence, some of it being splashed against the arch of the furnace, and some solidifying into most fantastic tree-like forms several inches in height. This behavior of Ag has been shown to be due to its property of mechanically absorbing O, at a temperature above its melting-point, which it gives off as it approaches the point of solidification, the escaping gas forcing up the crust of solid Ag formed upon the surface.

ful attraction for S, forming silver sulphide.* Silver spoons and door-knobs are tarnished by the minute quantities of H_2S present in the air.† The best solvent of Ag is HNO_3. The test of Ag in solution is HCl, which forms a cloudy precipitate of silver chloride. A solution of silver coin is blue, from the Cu it contains. Standard silver is whitened by being heated until the O of the air has converted a little of the Cu on the outside into CuO, which is dissolved by immersing in dilute H_2SO_4 or H_3N. The film of nearly pure Ag which then remains at the surface exhibits a want of lustre and is called *dead* or *frosted silver*. It is brightened by burnishing.

Compounds. — *Silver Nitrate,* $AgNO_3$, is sold in small, round sticks as *lunar caustic,* used as a cautery. It stains the skin and all organic matter black, especially when exposed to the light, owing to the formation of silver oxide, Ag_2O.‡ Hair-dyes and indelible inks contain $AgNO_3$. It is also the basis of photography (light-drawing) and daguerreotyping,§ which are both founded

* The perspiration from our bodies contains more or less S, and this, as it passes through our pockets, combines with any silver we may chance to have there.

† Those who have visited sulphur springs know the propriety of carefully protecting their watches, and of never wearing gold ornaments to the hot baths. Ag_2S is very easily dissolved by a little *dilute* ammonia (1 part of HN_3 to 20 of H_2O), which is therefore used for cleaning silver door-knobs.—*Oxidized silver*, as it is erroneously called, is made by immersing articles of silver in a solution obtained by boiling sulphur with potash, when the metal becomes coated with a thin film of sulphuret of silver.—BLOXAM.

‡ A very pretty experiment, illustrating the formation of this oxide, is performed by dropping into a test-tube of H_2O a few drops of silver nitrate in solution, and then adding potash, when a copious precipitate of the brown hydrate of Ag_2O will fill the tube. Now put in a little H_3N, which will instantly dissolve the silver oxide, and leave the liquid as clear and sparkling as spring-water.—The stain of silver nitrate may be removed by a strong solution of potassium iodide or the poisonous potassium cyanide.

§ The daguerreotype is named from M. Daguerre, the discoverer, who received a pension of 6,000 francs per year from the French government. A plate of Cu,

upon essentially the same principles. The general outlines of the photographic process are as follows: 1. Iodized collodion* is poured upon a clean glass plate, which, on evaporation, it covers with a transparent film. 2. The plate is put in the "nitrate of silver bath," † where the salt of silver is absorbed by the collodion film and changed to brom-iodide of silver. The plate is now ready for the picture. After the sitting, the plate is taken, carefully protected from the light, to the operator's room. Here the picture is, 3, *developed* by a solution of ferrous sulphate (protosulphate of iron) or pyrogallic acid (see p. 212): at the right stage the liquid is washed off, and the operation checked. 4. It is *fixed* with a solution of sodium hyposulphite, which dissolves the unaltered brom-iodide of silver. 5. It is washed, dried, and coated with amber varnish to preserve the film from accidental injury. The "*negative*" is now completed, and is a correct like-

plated on one side with Ag, is exposed to the vapor of I and Br until a compound of brom-iodide of silver is formed upon the surface. This is extremely sensitive to the light, hence the process is always conducted in a dark closet. The plate is then quickly carried, carefully covered, to the camera, and placed in the focus, where the rays of light from the person whose "picture is being taken" fall directly upon it. These rays decompose the brom-iodide of silver. The amount of this change is directly proportional to the number of rays that are reflected from different parts of the person to form the image in the camera. A white garment reflects all the light that falls upon it, so the corresponding part of the plate will be very much changed. A black garment reflects no light, so that part will not be changed at all. The different colors and shades reflect varying proportions of light, and so influence the plate correspondingly. When the plate is taken out of the camera, it is carefully covered again and carried quickly into the dark closet. No change can be detected by the eye; but on exposure to the vapor of Hg, wherever the Ag has been freed, the Hg will combine with it, forming a whitish amalgam, but it has no effect on the rest of the plate. The picture thus treated comes forth nearly perfect in its lights and shades. The undecomposed brom-iodide of silver is removed by a solution of sodium hyposulphite. A solution of gold chloride and sodium hyposulphite is then poured upon the plate and warmed. This golden varnish finishes the picture.

* Iodized collodion is composed of gun-cotton dissolved in alcohol and ether, to which are added ammonium iodide and cadmium bromide, or similar salts.

† The nitrate of silver bath contains nitrate of silver and iodide of silver in solution, and is acidulated with nitric acid.

ness, only the lights and shades are reversed. From this the pictures are, 1, "printed" by placing the negative upon a sheet of prepared paper,* and exposing it to the sun's rays. When the colors are sufficiently deepened, the picture is, 2, *toned* in the "toning-bath," which contains a little "bicarbonate of soda" and a minute quantity of gold chloride; 3, *fixed*, by sodium hyposulphite which dissolves the unaltered AgCl; 4, thoroughly washed in water frequently renewed; and, lastly, dried and mounted on card-board. The thoroughness of the third and fourth processes has much to do with the permanence of the picture. If any of the chloride or the compound formed by the hyposulphite be left, it will cause fading or discoloration.

PLATINUM.

Symbol, Pt....Atomic Weight, 197....Specific Gravity, 21.53. Fusing Point, about 4591° F. (?)

Source.—Pt † is chiefly found in the Ural Mountains, where it occurs in alluvial deposits, usually in small, flattened grains.‡

Preparation.—The "ore," as it is called, is separated from the earthy particles by washing. The grains of Pt remain behind with particles of Au, Fe_3O_4, and an alloy of Os and Ir. § The Au is removed by amalgamation, and the Fe

* This paper is "sensitized" by floating it on a solution of sodium chloride, and then on one of silver nitrate, thus filling the pores of the paper with the silver chloride, which is extremely sensitive to light.

† The word *platinum* signifies "little silver."

‡ The largest nugget ever found weighed 18 lbs.

§ Ir is named from Iris, the rainbow, because of the beautiful color of its salts

by a magnet. The Pt is then dissolved by melted Pb and afterward recovered from this alloy by cupellation.

Properties.—Pt resembles Ag in its appearance. It is one of the most ductile metals, wire being made from it so fine as to be invisible to the naked eye.* It is soluble in aqua-regia, but not in the simple acids. It does not oxidize in the air, is the most infusible of metals, and can be melted only by the heat of the compound blow-pipe or voltaic battery. In the arts it is fused in the former manner. These properties fit it for use as crucibles, and for this purpose it is invaluable to the chemist.

MERCURY.

Symbol, Hg....Atomic Weight, 200....Specific Gravity, 13.5.
Melting (Freezing) Point, – 39°F...Boiling Point, 662° F.

MERCURY is also called quicksilver, because it runs about as if it were alive, and was supposed by the alchemists to contain silver. It was known very anciently, and the mines of Spain were worked by the Romans.

Source.—*Cinnabar,* HgS, a brilliant red ore, is the principal source of this metal.† When sublimed with S, Hg forms the pigment known as "vermilion."

in solution. When combined with Os, it makes "iridosmine," used for the nibs of gold pens.

* Wollaston's Method, as it is called, consists in covering fine platinum wire with several times its weight of Ag, and then drawing this through the plates used for drawing wire until the finest hole is reached, when the wire is placed in HNO_3, which dissolves the Ag and leaves the Pt intact. This, in the form of the finest wire known, may be found in the solution by means of a microscope. (See *Physics*, p. 19.)

† Hg is found native in Mexico in very small quantities, where the mines are said to have been discovered by a slave, who, in climbing a mountain, came to a very steep ascent. To aid him in surmounting this, he tried to draw himself

Preparation.—Hg is readily prepared by roasting HgS in the open air. The S passes off as SO_2, while the Hg volatilizes and is condensed in earthen pipes.

Properties.—Hg emits a vapor at all temperatures above 40° F. Its solvent is HNO_3. It forms an amalgam * with gold or silver. This is its most singular property. A gold leaf dropped upon mercury disappears like a snowflake in water. Particles of Ag or Au, too fine to be seen by the eye, will be found by Hg and gathered from a mass of ore.

Uses.—Hg is extensively employed in the manufacture of thermometers and barometers; for silvering mirrors; † and for extracting the precious metals from their ores.

up by a bush which grew in a crevice above. The shrub, however, giving way, was torn up by the roots, and a tiny stream, of what seemed liquid silver, trickled down upon him.

* "Several years ago, while lecturing upon chemistry before a class of ladies, we had occasion to purify some quicksilver by forcing it through chamois skin. The scrap of leather remained upon the table after the lecture, and an old lady, thinking it would be very nice to wrap her gold spectacles in, accordingly appropriated it to this purpose. The next morning she came to us in great alarm, stating that the gold had mysteriously disappeared, and nothing was left in the parcel but the glasses. Sure enough, the metal remaining in the pores of the leather had amalgamated with the gold, and, entering, destroyed the spectacles. It was a mystery, however, which we could never explain to her satisfaction."—J. R. NICHOLS *in Fireside Science.*

† Mirrors were anciently made of steel or silver, highly polished. They were very liable to rust and tarnish, and so a piece of sponge, sprinkled with pumice-stone, was suspended from the handle for rubbing the mirror before use. Seneca, in lamenting over the extravagance of his time among the old Romans, says: "Every young woman now-a-days must have a silver mirror." The process of silvering ordinary mirrors is briefly as follows: Tin-foil is first spread evenly upon a marble table, and then the Hg is carefully poured over it. The two metals combine, forming a bright amalgam. A clean, dry plate of glass is then carefully pushed forward over the table so as to carry the superfluous Hg before it, and also prevent the air from getting between the glass and the amalgam. Weights are afterwards added to cause the film to cling more closely. In twenty-four hours the plate is removed, and in three or four weeks is dry enough to be framed. When we look in a mirror we rarely realize what it has cost others to thus minister to our comfort. The workmen are short-lived. A paralysis sometimes attacks them within a few weeks after they enter the manufactory, and it is thought remarkable if a man escapes for a year or two. Its effects are similar to those of calomel; the patient dances instead of walking, and cannot direct the motion of his arms, nor in some cases even masticate his food.

The action of Hg on the human system is too well known to need description. "In its metallic state, Hg has been taken with impunity in quantities of a pound weight" (*American Cyclopedia*), but when finely divided, as in vapor, mercurial ointment,* or "blue-pill," its effects are marked. It renders the patient extremely susceptible to colds; acts, as is generally thought, upon the liver, increasing the secretion of bile, and repeated doses produce "salivation."

Compounds.—*Mercuric Oxide,* HgO, "red precipitate," is interesting, as the substance from which Priestley discovered O gas.

Mercurous Chloride, HgCl, *Calomel,* is a white powder used in medicine. It can be easily distinguished from corrosive sublimate, since it is insoluble in H_2O, and hence, tasteless.

Mercuric Chloride, $HgCl_2$, *Corrosive Sublimate,* is a heavy, white solid, soluble in H_2O, and with a burning metallic taste. It has powerful antiseptic properties, and is used to preserve specimens in natural history. It is a deadly poison. Its antidote is white of eggs, milk, etc.

THE ALLOYS.

THESE are very numerous, and many of them possess properties so different from their elements that they almost seem like new metals. The color and hardness are changed, and sometimes the melting point is below

* This is vulgarly called "anguintum," which may be a corruption of the Latin term unguentum (unguent). It is used in cutaneous diseases.

that of any one of the constituents. The proportions of the metals used vary. The following is a fair average:

Type-metal contains 3 parts of lead* to 1 of antimony.†

Pewter contains 4 parts of Sn and 1 of Pb.

Britannia consists of 100 parts of Sn, 8 of Sb, 2 of Bi,‡ and 2 of Cu.

Brass is 2 parts of Cu and 1 of Zn.

German Silver contains Cu, Zn, and Ni § (brass whitened by nickel).

Soft Solder, used by tinsmiths, is made by melting Pb and Sn together, the usual proportion being half-and-half.

Hard Solder is composed of Cu and Zn.

Fusible Metal melts at 201°, and spoons made of it will fuse in hot tea. It can be melted in a paper-crucible over a candle. It consists of Bi, Pb, and Sn. Yet

* An improved kind lately introduced is 2 parts Pb, 1 part Sn, and 1 Sb.

† Antimony was discovered by Basil Valentine, a monk of Germany, in the fifteenth century. It is said that, to test its properties, he first fed it to the swine kept at the convent, and found that they thrived upon it. He then tried it upon his fellow-monks, but perceiving that they died in consequence, he forthwith named the new metal, in honor of this fact, *anti-moine* (anti-monk), whence our term *antimony* is derived. It is a brittle, bluish-white metal, with a beautiful laminated, star-like, crystalline structure. It is used simply as an alloy for type-metal, Britannia-ware, etc. Its test is H_2S, which throws down a brilliant orange-colored precipitate. Melt a small fragment of Sb before the blow-pipe, and throw the melted globule upon an inclined plane. It will instantly dart off in minute spheres, each followed by a long trail of smoke.

‡ Bismuth is a reddish-white metal used only in alloys and in the construction of thermo-electric piles. (See *Physics*, p. 249.)

§ Ni, like Co, is a constituent of meteorites. It is mined in Pennsylvania for the United States government to make into cents. Formerly its principal use was in German silver, but of late it has been employed extensively in the manufacture of the best plated-ware. (See *Physics*, p. 238.) Its silvery whiteness, when pure, its high polish, which often lasts for years, and its hardness, almost equal to that of steel, eminently fit it for the plating of mathematical and other delicate instruments. The salts of Ni have a handsome green tint. The rare gem *chrysoprase* is colored by the oxide.

the first metal melts at 507°, the second at 617°, and the third at 442°.

Bronze is 95 parts of Cu, 4 of Sn, and 1 of Zn.

Gold is soldered with an alloy of itself and Ag; ***Silver,*** with itself and Cu; ***Copper,*** with itself and Zn: the principle being that the metal of lower fusing point causes the other to melt more easily.

Coin.—The precious metals, when pure, are too soft for common use. They are therefore hardened by other metals. The gold coin of the United States consists of 9 parts of gold and 1 of alloy. The alloy is composed of 9 parts of Cu, whitened by 1 of Ag, so as not to darken the gold coin. Silver coin is 9 parts of Ag and 1 of Cu. The nickel cent is 88 parts of Cu and 12 of Ni. Cu being cheaper than Ni, it is used to make the coin larger. The term *carat,* applied to the precious metals, means $\frac{1}{24}$ part. Therefore, gold 18 carats fine, contains $\frac{18}{24}$ of gold and $\frac{6}{24}$ of alloy.

Shot is an alloy of about 1 part of As to 100 of Pb. The manufacture is carried on in what are called "shot-towers," some of which are two hundred and fifty feet high. The alloy is melted at the top of the building, and poured through colanders. The metal, in falling, breaks up into drops, which take the spherical form (see *Physics,* pages 44 and 192), harden, and are caught at the bottom in a well of water which cools the shot and also prevents their being bruised in striking. The shot are dipped out, dried, and then assorted, by sifting in a revolving cylinder, which is set slightly inclined and perforated with holes, increasing in size from the top to the bottom. The shot being poured in at the top, the small ones drop through first, next the larger, and so on, till the largest reach the bottom. Each size is received in its own box. Shot are polished by being agitated for several hours with black-lead, in a rapidly revolving wheel. They are finally tested by rolling them down a series of inclined planes placed at a little distance from each other. The spherical shot will jump from one plane to the next, while the imperfect ones will fall short, and drop below; or sometimes, by rolling down a single inclined plane, the spherical ones will go to the bottom, while the imperfect ones roll off at the sides.

Oreide is a beautiful alloy of brass resembling gold, but it soon tarnishes by exposure to the air.

Aluminum Bronze, or gold, is an alloy of Al and Cu. It is elastic, malleable, and very light. It strikingly resembles gold, and is much used instead of that costly metal.

REVIEW OF THE PROPERTIES OF THE METALS.

Oxidation.—K and Na have an intense attraction for O, while Au and Ag have little affinity for it, and are therefore found native.

Density. — L is the lightest liquid known. Pt is the heaviest solid, being over twenty-one times as heavy as H_2O, while K and Na will float upon it.

Melting Point.—Hg is liquid at all ordinary temperatures. K and Na melt beneath the boiling point of H_2O; Zn below a red heat, and Cu above; Co, Ni, and wrought-iron require the greatest heat of the forge (4000°), while Pt and Os melt only in the flame of the oxy-hydrogen blow-pipe. Sn melts at the lowest temperature (442°) of any of the ordinary metals.

Color.—The most common color is white, of varying shades. It is nearly pure in Ag, Pt, Cd, and Mg; yellowish in Sn; bluish in Zn and Pb; gray in Fe, and reddish in Bi. Cu is a full red, and Au a bright yellow.

Malleability.—Au, Ag, and Cu are the most malleable of the metals; Au, Ag, and Pt are the most ductile.

Brittleness.—Sb and Bi may be easily powdered; Zn may be broken with more difficulty, while the fibrous metals are exceedingly tough.

Tenacity.—Steel is the most, and lead the least, tenacious of the metals; the proportion being as 1 to 42.

Special Properties.—Certain of the metals are valuable because of their peculiar properties. Thus, Hg, because it will form an amalgam, and is a liquid at all ordinary temperatures; Sb, because it hardens Pb and Sn; Bi and Cd, because they render Pb and Sn more easily melted; Ni, because it whitens Cu; Mg, for its brilliant light; Au, for its rarity and lustre; Fe, for the diverse properties it can assume in wrought and cast iron, and in steel, and because it is the only metal which can be used for the magnetic needle and electro-magnet; Cu, for its ductility and its conductibility of electricity; and Pt, for its infusibility.

PRACTICAL QUESTIONS.

1. Pb is softer than Fe; why is it not more malleable?
2. What is the cause of the changing color often seen in the scum on standing water?
3. How can the spectra of the metals be obtained?
4. Ought cannon, car-axles, etc., to be used until they break or wear out?

5. Why is "chilled iron" used for safes?

6. Does a blacksmith plunge his work into water merely to cool it?

7. What causes the white coating made when we spill water on zinc?

8. Is it well to scald pickles, make sweetmeats, or fry cakes in a brass kettle?

9. What danger is there in the use of lead pipes? Is a lining of Zn or Sn a protection?

10. Is water which has stood in a metal-lined ice-pitcher healthful?

11. If you ask for "cobalt" at a drug-store, what will you get? If for "arsenic?"

12. What two elements are fluid at ordinary temperatures?

13. Should we touch a gold ring to mercury? *

14. Why does silver blacken if handled?

15. Why does silver tarnish rapidly where coal is used for fires?

16. Why is a solution of a silver coin blue?

17. Why will a solution of silver nitrate curdle brine?

18. Why does writing with indelible ink turn black when exposed to the sun, or to a hot iron?

19. What alloys resemble gold?

20. Why does a fish-hook "rust out" the line to which it is fastened?

21. Why do the nails in clap-boards loosen?

22. Show that the earth's crust is mainly composed of burnt metals.

23. What kind of iron is used for a magnet? For a magnetic needle?

24. Why does a tin pail so quickly rust out when once the tin is worn through?

25. Why is the zinc oxide found in New Jersey red, when zinc rust is white?

26. Should we filter a solution of permanganate of potash through paper?

27. Why are wood, cordage, etc., sometimes soaked in a solution of corrosive sublimate?

* If the surface is only whitened, the Hg may be removed with dilute HNO_3, and the ring be polished to look as before. The Hg will soon penetrate the gold and render it brittle.

28. Why does the white paint around a sink turn black?

29. Why is aluminum, rather than platinum, used for making the smallest weights?

30. How would you detect the presence of iron particles in black sand?

31. Which metals can be welded?

32. When the glassy slag from a blast-furnace has a dark color, what does it show?

33. In welding iron the surfaces to be joined are sometimes sprinkled with sand. Explain.

34. What is the difference between an alloy and an amalgam?

35. Steel articles are blued to protect from rusting, by heating in a sand-bath. Explain.

36. Give the rational formulæ for copperas and white lead.

37. Why is Hg used for filling thermometers?

38. What oxide is formed by the combustion of Na, K, Zn, S, Fe, Pb, Cu, P, etc.? Which are bases? Acids? Give the common name of each.

39. Is charcoal lighter than H_2O?

40. Name the vitriols.

41. Is Mg a monad or a dyad? Zn?

42. Name some dibasic acid.

43. Name a neutral salt. An acid salt.

44. Calculate the percentage of water contained in crystallized copper sulphate. Sodium sulphate. Calcium sulphate. Alum.

45. What is the test for Ag? Cu?

46. What weight of crystallized "tin salts" ($SnCl_2, 2H_2O$) can be prepared from one ton of metallic tin?

47. 100 parts by weight of silver yield 132.8+ parts of silver chloride. Given the combining weight of chlorine, required that of silver.

48. What is the composition of slaked lime?

49. How is ferrous sulphate obtained? How many tons of crystals can be obtained by the slow oxidation of 230 tons of iron pyrites containing 37.5 per cent. of sulphur?

50. Required 500 tons of soda crystals; what will be the weight of salt and pure sulphuric acid needed?

51. Describe the uses of lime in agriculture.

52. How many tons of oil of vitriol, containing 70 per cent. of pure acid (H_2SO_4), can be prepared from 250 tons of iron pyrites, containing 42 per cent. of sulphur?

III.

Organic Chemistry.

"Thus the Seer,
With vision clear,
Sees forms appear and disappear,
In the perpetual round of strange,
Mysterious change
From birth to death, from death to birth,
From earth to heaven, from heaven to earth;
Till glimpses more sublime
Of things, unseen before,
Unto his wondering eyes reveal
The Universe as an immeasurable wheel
Turning forevermore
In the rapid and rushing river of Time."

LONGFELLOW

ORGANIC CHEMISTRY.

INTRODUCTION.

Necessity of Organization.—We have thus far spoken of the various elements of matter. We have found many which are necessary to the growth of our bodies, but still we cannot live upon them. We need phosphorus, but we cannot eat it, for it is a deadly poison. We need Fe, but it would make a most unsavory diet. We need CaO, but it would corrode our flesh. We need H, but it must be combined with O as H_2O to be of any value to us. We need C, but charcoal would form a very indigestible food. If we were shut up in a room with all the elements of nature, we not only could not combine them so as to produce those organic substances necessary to our life and comfort, but we should actually die of starvation. We thus find that the mineral matter must be organized in some manner before we can use it to advantage.

Plants Organize Matter.—We have seen that in the plant the sunbeam decomposes the poisonous CO_2 and furnishes us the life-giving O; that we cannot create force ourselves, or draw it direct from the sun, but must take that which the plant has hoarded for us. We shall now find that, in addition, the plant changes *inorganic* matter to *organic*. It takes up the elements we need for our growth and for use in the arts, and combines them into plant-products, such as wood, starch, sugar, etc.—We are thus dependent upon the vegetable world for the grand staples of commerce and of luxury—all that we eat, drink, or wear. Each tiny leaf, every tree and shrub, every spire of grass even, is working constantly for us. The earth was once a burnt body—the cinders of the vast fire amid which it had its origin. (See *Geology*, page 17.) Every organized substance now on its surface has been rescued from the grasp of O by the plants.

Difference between Organic and Inorganic Bodies.—1. While inorganic bodies deal with sixty-three elements, organic are composed principally of only four, C, H, O, and N. As C is their characteristic element, they are frequently styled the "carbon compounds." 2. While inorganic molecules consist of only a few atoms, and are therefore very simple in their construction, as: H_2O, CO_2, K_2O, organic frequently contain a large number, and are extremely complex, as: Sugar = $C_{12}H_{22}O_{11}$, having 45 atoms in a molecule; stearin = $C_{57}H_{110}O_6$, having 173 atoms; albumen = $C_{72}H_{110}N_{18}SO_{22}$, having 222 atoms, and even more, according to some authorities. 3. While inorganic bodies are formed and remain fixed in one state under the influence of chemical affinity, organic grow rapidly, change constantly, and when life ceases, as rapidly decay, and are transformed into inorganic substances. 4. Owing to their complex structure, and the presence in many of them of the negative N, they form most unstable compounds. In this we find the cause of their quick decay. The vital principle alone holds them together, frequently in opposition to the laws of chemical affinity; and the instant that is removed, the tendency is to seek new affinities and form new compounds. On the other hand, inorganic are generally burnt bodies. Their chemical affinities are satisfied, and hence at rest.

The Number of Carbon Compounds greatly exceeds that of all the other elements combined, and is constantly increasing. The labor of modern chemists is largely devoted to the subject, and the field opens and broadens with every discovery. The methods of classification are unsettled,* and new and conflicting theories yet contend on this border-ground of chemical knowledge.

Isomerism.—Isomeric compounds are those which consist of the same elements in the same proportion.—*Example:* Sugar and gum-arabic have the same molecular formula, $C_{12}H_{22}O_{11}$.—GÉLIS. The difference between such compounds has been supposed to lie in a dissimilar grouping of the atoms about one another, as the same pieces upon a checker-board may be variously arranged; or as the letters p-l-e-a may also spell l-e-a-p, or p-e-a-l, or p-a-l-e: yet nothing is definitely known.

Allotropism.—The individual elements are also susceptible of allotropic states; as, for instance, the C in a compound may be in any one of its three allotropic forms. These two principles of isomerism and allotropism run through organic chemistry, and account, in some measure, for the immense number of its compounds. Still, one

* Even Miller, in his great work on the Elements of Chemistry, which embodies the best results of modern research, uses the same division as the older authorities, and nearly the same as that which was adopted in the former edition of this work.

can hardly understand how olefiant gas and india-rubber, the fragrance of a rose and the odor of a kerosene-lamp, should consist of the same elements, C and H, only in varying proportions.

STARCH, WOODY FIBRE, AND SUGAR.

1. STARCH.

$(C_6H_{10}O_5.)$

Fig. 68.

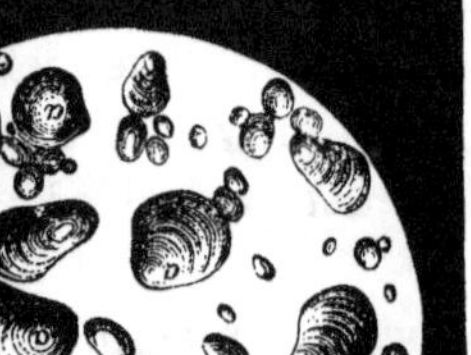

Potato Starch.

Source. — Plants accumulate it, 1, in their roots, as the carrot, the turnip, etc.; 2, in subterranean stems, as the potato, of which it forms about 20 per cent.; 3, in the base of their leaves, as the onion; 4, in the seed, as corn, of which it forms about 80 per cent.; 5, in the embryo, as the bean, the pea, etc. In all these it is stored up for the future growth of the plant. It is kept in its starch form (lest it

Fig 69.

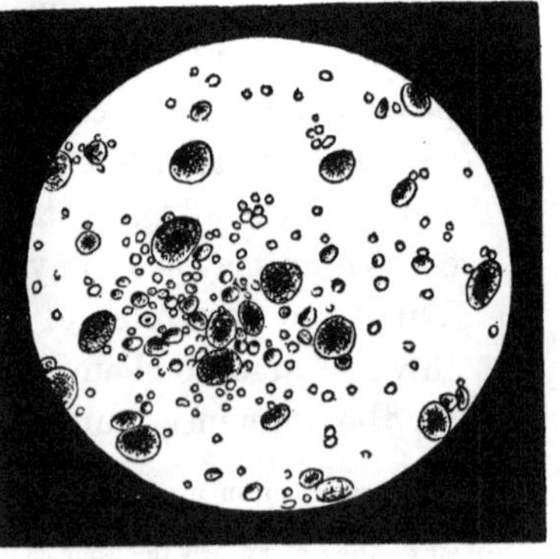

Wheat Starch.

dissolve in the first rain), and then turned to sugar only when the plant needs it in growing. (See p. 194.) Under the microscope, each vegetable is found to have its peculiar form of starch granule, so that in this way any adulteration is easily detected.*

Preparation.—Starch is made from wheat, corn, potatoes, etc. The process is essentially the same in all. The potato, for example, is ground to a pulp, and then washed with cold water. The starch settles from this milky mass as a fine, white precipitate.

Properties.— Starch is insoluble in cold water; in hot, it absorbs H_2O, swells, and the granules burst, forming a jelly-like liquid, used for *starching*. The swelling of rice, beans, etc., when cooked, is owing to this property. By heating to 400° when dry, starch undergoes a peculiar change into a substance known as dextrine,† used as a mucilage on envelopes and adhesive stamps, for making "fig-paste," and stiffening chintzes. The test of starch is I, which forms in solution the blue iodide of starch. Sago is the starch from the pith of the palm-tree; tapioca and

Fig. 70.

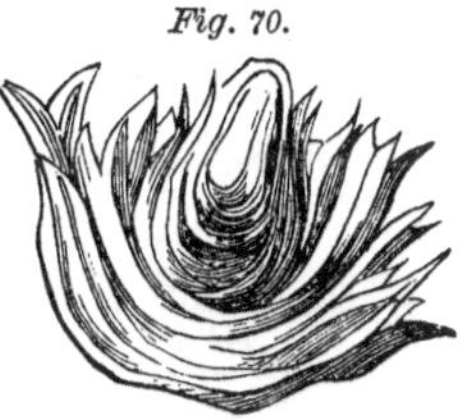

Bursting of Starch Granule.

* "The structure of the grains of starch is very beautifully displayed by placing some of them in contact with a drop of concentrated solution of zinc chloride (tinged with a little free iodine) on the field of the microscope. No change takes place in the granules until a little water is added. They then become of a deep blue color, and gradually expand; at first a frill-like plaited margin is developed around the globule; by degrees this opens out; the plaits upon the globule may then be seen slowly unfolding, and may be traced in many cases into the wrinkles of the frill; ultimately the granules swell up to twenty or thirty times their original bulk, and present the appearance of a flaccid sac."—BUSK.

† Dextrine is isomeric with starch, but is not discolored by I.

arrow-root are made from the roots of South American marshy plants.*

GUM is found in the juices of nearly all plants, and frequently exudes, as in the peach, plum, and cherry. It is soluble in water, but not in alcohol. *Gum-arabic,* which flows in transparent tears from an acacia tree, is the purest form.† *Mucilage,* which occurs in gum tragacanth, linseed, quince-seed, etc., is a modification of gum, and is insoluble in H_2O. It forms with it, however, a gelatinous liquid, which is exceedingly useful.

Vegetable Jelly.—A variety of gum called *pectose* exists in nearly all fruits and vegetables. It gives to them their hardness while green.—FREMY. In the process of ripening, or by heat, acids, etc., it is turned into *pectin.* We find this abundant in the thick juice which exudes from an apple while baking. In the making of jellies, pectose is converted into a mixture of pectosic and pectic acids.

---

2. WOODY FIBRE ($C_6H_{10}O_5$).

Sources.—If a thin slice of wood be examined under the microscope, it will be seen to consist of a fibrous substance incrusted and compacted with woody matter. The former is called *cellulin* ($C_6H_{10}O_5$).‡ It composes the cells

* Very many of the farinaceous preparations sold for the sick and invalid, under high-sounding names, are simply wheat or corn starch.

† It is a soluble salt, being composed of arabic acid ($C_{12}H_{22}O_{11}$, *Gélis*), combined with K and Ca.

‡ It is probable that the molecule of woody fibre is some multiple of this formula, as $C_{18}H_{30}O_{15}$.

of all plants, giving them strength and firmness, and is found even in delicate fruits, holding their luscious juices. It occurs in various modifications, in wood, nut-shells, and fruit-stones. In the heart of a tree, its cells are hard and dense; in the outer part, they are soft and porous; in elder-pith and cork, light and spongy; in flax and cotton, long, pliable, and fibrous; in the bran of wheat and corn, digestible.

Secretion.—All vegetation consists of these simple cells. They seem alike to the eye, yet they have a very diverse power of secretion. The cell of the sugar-maple converts its sap into sugar; the milk-weed, into a milky juice; the caoutchouc, into rubber; the rhubarb-plant, into oxalic acid; and the rose-petal, into the most delicate of perfumes.

Cells are always true to themselves. There seems to be a law of God stamped on each one, so that when we cut a tiny bud from one tree and graft it into another, it remains consistent with itself. It develops into a limb, and years pass by. The few single cells become a myriad, yet they have not changed. The sap flows upward in the tree; but at a certain point—a hidden threshold which no human eye can discern, it comes under a new and strange influence. Here it is transformed, and produces fruit and flowers, in accordance with another and different growth. Somehow quince-juice is made into pears, locust-juice blooms out into fragrant acacias, and sweet and sour apples hang upon the same limb.

Uses.—These are wonderfully various. Woody fibre is woven into cloth, built into houses, twisted into rope, twine, and thread, pressed into paper, cut into fuel, carved into furniture. We eat it, wear it, walk on it, write on it, sit on it, print on it, pack our clothes in it, sleep in it, ride in it, and burn it.

PAPER is made from cotton, linen, straw, or any substance containing cellular tissue. The finest writing-paper is manufactured from linen rags. These are first "shredded" upon scythe-blades—*i. e.*, the seams are ripped open, buttons cut off, and the dust shaken out. 2d. They are steamed in a solution of chloride of lime for ten or twelve hours until they are thoroughly bleached. 3d. They are received by a machine that alternately lacerates them by a cylinder set with razor-like blades, and washes them with pure, cold water for six hours, until they are reduced to a mass resembling rice and milk. 4th. This pulp receives a delicate blue tint from *smalt.** 5th. It is diluted with H_2O nearly to the consistency of milk, and strained to remove the waxed ends and knots of thread that cause the little lumps which catch our pen when we write rapidly on poor paper. 6th. It flows over an endless belt of wire-gauze, about thirty feet in length, through which the water steadily drips from the pulp, as it slowly passes along, gaining consistency and firmness. 7th. It comes to a part of the belt called the "dandy-roll," consisting of a cylinder, on the surface of which are wires arranged in parallel rows, or fancy letters, which print upon the moist paper a design—constituting what is termed "laid," or "wire-woven," paper. 8th. The paper, very soft and moist as yet, passes between rollers that squeeze out the water; then between others which are hot, and dry it. 9th. It comes to a vat of sizing, composed of glue and alum, into which it plunges, and at the

* Powdered glass colored with oxide of cobalt.

opposite side emerges only to go between other rollers that press and dry it—at the end of which it passes under a cylinder, set with knives that clip the roll into sheets of any desired size.

Paper Parchment is prepared by plunging unsized paper for a few seconds in H_2SO_4 of a specified strength, then washing off the acid. This, in some unknown way, changes its appearance and character, so that it resembles parchment, while its toughness is five times that of the paper from which it was made.

Linen is made from the inner bark of flax. The plant is first *pulled* from the ground to preserve the entire length of the stalk; next "rotted" by exposure to air and moisture, when the decayed outer bark is removed by "breaking;" then, by "hatcheling," the long, fine fibres are divided into shreds, and laid parallel, while the tangled ones are separated as "tow." It is then bleached on the grass, which renders the gray coloring-matter soluble by boiling in lye. The whitened flax is lastly woven into cloth.

COTTON consists of the beautiful hollow, white hairs arranged around the seed of the cotton-plant. As it is always pure and white—except Nankin cotton, which is yellow—it would require no bleaching did it not become soiled in the process of spinning, etc.

Pyroxylin (*pur*, fire, and *xulon*, wood), *Gun-Cotton*, is prepared by dipping cellular tissue—cotton, saw-dust, printing-paper, etc.—in a mixture of HNO_3 and H_2SO_4 of a certain specific gravity. It is then carefully washed and dried. It is not materially changed in appearance, but a part of its H has been replaced by NO_2, and it has become very inflammable. It will burn at a temperature more than 200° below that required to ignite gunpowder, while its explosive force is much greater.

Collodion is a solution of gun-cotton in sulphuric ether and alcohol. It forms a syrupy liquid, which is much used by photographers.

3. SUGAR.

CANE-SUGAR ($C_{12}H_{22}O_{11}$),* *Sucrose*, is obtained from the sap of the sugar-maple, and the juice of the sugar-cane, sorghum, and beet. In making it from the sugar-cane, the canes are crushed between iron cylinders, to express the juice. As the liquid sours very soon, from the heat of the climate in which it grows, a little lime is added to neutralize the acid, and it is then evaporated to a thick jelly, and set aside to cool. The sugar crystallizes readily, forming *brown sugar*, which is put in perforated casks to drain. The drainings, "mother-liquor," constitute molasses.

Refining.—Brown sugar is dissolved in H_2O, filtered through twilled cotton to remove the coarse impurities, and then through a deep layer of animal charcoal. The colorless solution is next evaporated in vacuum pans from which the air is exhausted, so that the sugar boils at so low a temperature as to avoid all danger of burning. When sufficiently concentrated, the liquid is removed and set aside to crystallize. If the mass of crystals is dried in moulds, it forms *loaf sugar;* if in centrifugal machines,

* A very brilliant experiment showing the presence of C in $C_{12}H_{22}O_{11}$ is obtained by putting on a clean, white plate, a mixture of finely-pulverized white sugar and $KClO_3$. Upon adding a few drops of H_2SO_4, a vivid combustion will ensue. By mixing with the sugar a few iron and steel filings, and performing the experiment in a dark room, or out of doors at night, fiery rosettes will flash through a rose-colored flame, and produce a fine effect.

*granulated sugar.** The drainings constitute "syrup," "sugar-house molasses," etc.

Confectionery.—Terra alba (white earth), is imported from Ireland for use in lozenges, drops, etc.† Confectionery is often colored by dangerous poisons, so that prudence would forbid the use of any colored candy. Licorice drops are frequently only the poorest brown sugar, terra alba, and a flavoring of licorice to make the unwholesome mixture palatable. Gum-drops are made, not from gum-arabic, but generally of a species of glue manufactured out of hoofs, parings of hides, etc. However repugnant it may appear, this substance is perfectly clean and wholesome. Rock candy is formed by suspending threads in a strong solution of sugar. It crystallizes upon the rough surface in large, six-sided prisms.

Caramel, familiarly called burnt sugar, is formed whenever sugar is heated to about 420°, as when sweetmeats boil over on the stove; H_2O is lost and C remains in excess. It is used by confectioners and for coloring liquors.

GRAPE-SUGAR ($C_6H_{12}O_6$), *Dextrose,* is found in honey, figs, and many kinds of fruit. Its sweetening power is about two-fifths that of cane-sugar.

Sugar from Starch.—The difference in the constitution of starch and grape-sugar is only H_2O. By

* This apparatus consists of a cylindrical drum mounted upon a vertical axis, to which a rapid rotary movement can be given. The outer side of this drum is made of a stout but closely-woven network. The drum is inclosed in a large, fixed, cylindrical vessel capable of holding the liquid which may pass out through the network. A charge of syrup is placed in the inner drum, which is then made to revolve rapidly. The syrup escapes through the wire-gauze into the outer drum while the crystals are rapidly dried.

† We can and should test all the candy we purchase by putting a small piece in a glass of water. Whatever settles to the bottom and remains undissolved is an adulteration.

slowly heating potato-starch with dilute H_2SO_4 it is transformed into a syrup, from which the dextrose will separate in crystals. The weight of the sugar will exceed that of the starch by the additional H_2O. The acid acts by catalysis, being itself unchanged in the process.*

"Candied Jellies, Preserves, etc."—The sugar of many kinds of ripe fruits consists of grape or cane sugar mixed with fruit sugar. The latter changes gradually into grape sugar, and crystallizes as in honey, dried figs, etc.†

FERMENTATION.

If a solution of starch or grape-sugar be exposed to the air it will undergo no change; but if there be added a little ferment,‡ or any albuminous substance (*i. e.*, one

* Saw-dust, paper, and even rags can in the same way be converted into sugar. Indeed, Prof. Pepper speaks of seeing some made out of an old shirt. Wonderful beyond our comprehension is that chemical force which can transform a cast-off garment into a substance which will delight the palate, or a snowy page on which thought may be inscribed. Thus the chemist faintly imitates nature, which, ever out of waste and refuse, springs afresh. The fair petals of the lily rest upon the black mud of the swamp, and the products of decay come back to us in objects of use and forms of beauty.

† Grape sugar is isomeric with fruit sugar, but is much sweeter. The former, as it is noted for its right-handed rotation of the plane of polarized light, is called *dextrose* (*dextra*, right), and the latter, from its left-handed rotation, lævulose (*lævus*, left). (See *Physics*, p. 170.)

‡ In many cases, spontaneous fermentation sets in without the apparent addition of any ferment: thus wine, beer, milk, etc., when allowed simply to stand exposed to the air, become sour, or otherwise decompose. These changes are, however, not effected without the presence of vegetable or animal life, and are true fermentations: the *sporules*, or seeds of these living bodies, always float about in the air, and on dropping into the liquid begin to propagate themselves, and in the act of growing evolve the products of the fermentation. If the above liquids be left only in contact with air which has been passed through a red-hot platinum tube, and thus the living sporules destroyed; or if the air be simply filtered by passing through cotton wool, and the sporules prevented from coming into the liquid, it is found that these fermentable liquids may be preserved for any length of time without undergoing the slightest change.—ROSCOE.

containing N), in a decomposing state, it will immediately commence breaking up into new compounds. The ferment acts by catalysis, its presence seeming to overcome the unstable equilibrium of the chemical forces, and causing the large molecules to drop into smaller ones. There are two stages in this chemical change.

1st. ***Alcoholic Fermentation.***—In this, *the grape-sugar is resolved into alcohol and carbonic anhydride.* The former remains in the liquid, while the latter escapes in little bubbles of gas. The reaction may be represented thus: $C_6H_{12}O_6 = 2C_2H_6O + 2CO_2$.

2d. ***Acetic Fermentation.****—The second stage succeeds the first immediately, if not checked, and by absorbing O from the air, *the alcohol is broken up into acetic acid and water;* thus, $C_2H_6O + O_2$ (from the air) $= C_2H_4O_2 + H_2O$.

Yeast is formed during the process of fermentation. It consists of microscopic plants (*Mycoderma cerevisiæ*) which increase by the formation of multitudes of tiny cells not more than $\frac{1}{250}$ of an inch in diameter. In the brewing of beer they spring up in great abundance, making common brewer's yeast.†

Malt.—In making malt, the barley is thoroughly soaked in water, and then spread on the floor of a dark room, to heat and sprout. Here a curious change ensues, identical with that which takes place in every planted seed. Each one contains starch and a nitrogenous sub-

* There are also other forms of fermentation, as the lactic, yielding lactic acid—the acid of sour milk; butyric, yielding butyric acid, etc.

† The yeast-cakes of the kitchen are formed by exposing moistened Indian meal, containing a ferment, to a moderate temperature, until the gluten or albuminous matter of the cake has undergone this alcoholic fermentation. They are then laid aside for use.

stance called *gluten*. The tiny plant not being able to support itself in the beginning, has here a little patrimony with which to start in life; but, as the starch is insoluble in the sap, it must first be changed to a soluble form. We see, therefore, the need of a ferment; but it would not answer to store up in the seed an active ferment, as that might cause a change before the plant was ready to grow, and thus the plant's capital be wasted. The gluten acts, therefore, as a *latent* ferment. As soon as the seed is planted it absorbs moisture from the ground, is turned into *diastase*—an *active* ferment*—the starch is converted into dextrine and sugar, dissolved, and immediately applied to the uses of the growing plant. This change takes place in the *malting-room*. The barley sprouts, and a part of its starch is turned to sugar, so as to give it a sweetish taste. If this germination were allowed to proceed, the little barley sprout would turn the sugar into woody fibre. To prevent this, the grain is heated in a kiln until the germ is destroyed. Barley in this condition is called *malt*, and is then transported to the breweries.

Brewing Beer.—The malt is crushed and digested in water, to convert the remaining starch into dextrine and sugar. Hops and yeast are added, and fermentation immediately commences. Bubbles of gas rise to the top with a low hissing sound, yeast gathers in a foamy cream that comes to the surface of the tub, while the alcohol gradually accumulates in the liquid. The beer is now drawn off into tight casks, where it undergoes a second

* Malt does not contain more than $\frac{1}{500}$ of its weight of diastase; one part of this substance being sufficient to change 2,000 parts of starch into dextrine and sugar.—PERSOZ AND PAYEN.

fermentation; the flavor ripens, and the CO_2 collecting, gives to the liquor, when drawn, its sparkling, foamy appearance.

Lager Beer (*Lagern,* to lie) is so called because it is allowed to lie for months in a cool cellar, where it ripens very gradually. It is also fermented much more slowly and perfectly than ale or porter.

Wine is made from the juice of the grape. The juice, or *must,* as it is called, is placed in vats in the cellar, where the low temperature produces a slow fermentation. When all the sugar is converted into alcohol and CO_2, a dry wine remains; when the fermentation is checked, a sweet wine is the result; and when bottled while the change is still going on, a brisk, effervescing wine, like champagne, is formed. The flavor or "bouquet" of wine is due to the slow formation of a fragrant and aromatic ether.* (See p. 204.) The tartaric acid of the grape gradually separates and collects on the sides and bottoms of the casks in a white incrustation—cream of tartar.

Alcohol in Beer, Wine, etc.—Alcohol is the intoxicating principle of all varieties of liquors, ale, beer, wine, cider, and the domestic wines. Ale and porter contain from 6 to 8 per cent. of alcohol; wine varies from 7 per cent. in the light claret to 17 per cent. in the strong Port and Madeira; brandy and whisky have from 40 to 50 per cent.

Ardent Spirits.—When any fermented liquor is distilled, the alcohol passes over, together with water and some fragrant substances which are condensed. In this way brandy is made from wine; rum, from fermented

* Œnanthic ether, a liquid with a powerful odor, which causes the peculiar smell of grape-wine.

molasses or cane-juice; whiskey, from fermented corn, rye, or potatoes; and gin, from fermented barley and rye, afterward redistilled with juniper-berries. The accompanying cut represents an apparatus used for this distilla-

Fig. 71.

A Still.

tion. A is the boiler, B the dome, C a tube passing into S, the condenser, where it is twisted into a spiral form called the worm, in which the vapor from the boiler is condensed, and drops out at D. (See *Physics,* page 188.)

Alcohol (C_2H_6O)* is prepared by distilling whiskey, and is sometimes called spirits of wine. It boils at 173°, and has never been frozen even at −166° F. It contains,

* Ethyl alcohol or vinic alcohol. It is also known as *ethyl hydrate.* (See *Marsh-gas Series*, p. 201.)

when strongest, 10 per cent. of H_2O, which can be separated by adding some substance like lime, which has a strong affinity for H_2O. It is then called *anhydrous* or absolute alcohol. When C_2H_6O is exposed to the air the spirit evaporates, while moisture is absorbed from the atmosphere.* It burns without smoke and with intense heat, owing to the abundance of H and deficiency of C, and is therefore of great value in the arts. It is also of incalculable importance as a solvent of many substances —roots, resins, fragrant oils, etc.

Effects of Alcohol.—When pure it is a deadly poison. When diluted, as in the ordinary liquors, it is stimulative and intoxicating. Its influence is on the brain and nervous system;—deadening the natural affections, dulling the intellectual operations and moral instincts; seeming to pervert and destroy all that is pure and holy in man, while it robs him of his highest attribute—reason. (See *Physiology*, p. 150.)

Ether [$(C_2H_5)_2O$].—*Sulphuric Ether* is formed by the distillation of C_2H_6O with H_2SO_4.† It has a fragrant odor, boils at about 94°, and burns with more light and smoke, but less heat, than alcohol. Its vapor is thirty-seven times heavier than H, and can be poured like CO_2.‡ By the action of different acids on C_2H_6O, other ethers are produced, viz., nitric, acetic, butyric,§ etc.

* The chemist discovers this when he neglects to put the extinguisher on his alcohol-lamp, and finds that he cannot relight it without moistening the wick with fresh alcohol.

† Though thus named, this ether contains no S. It is known as *ethyl oxide.* (See page 202.)

‡ It should be used with care in the presence of a light, as it is inflammable and becomes explosive when mixed in proper proportions with air.

§ This has the odor of pine-apple, and is sold as pine-apple oil. The melon and strawberry are supposed to owe a portion of their flavor to this ether.

Chloroform ($CHCl_3$) is made by distilling C_2H_6O with chloride of lime. It is colorless, volatile, of a sweet taste, and should be free from any unpleasant odor when evaporated on the hand. It is used as a solvent of I, P, S, and caoutchouc, and as an anæsthetic. The value of ether and chloroform in alleviating pain, is beyond estimate.

Chloral (C_2Cl_3HO) is formed by passing Cl through absolute alcohol. It is an oily liquid which combines with H_2O, making *Chloral Hydrate*, a white, crystalline substance, much used to induce sleep. Taken in proper quantities it is entirely safe, and is exceedingly pleasant in its influence.

Acetic Acid ($C_2H_4O_2$, *acetum*, vinegar) forms from two to four per cent. of common vinegar, whence its name. The strongest acetic acid is known as the *glacial*, since it crystallizes into an ice-like solid at 63°. It has an aromatic taste and pungent odor, and, after a time, blisters the skin.

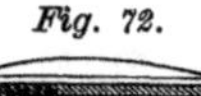
Fig. 72.

Making Vinegar.

Preparation.—Vinegar is made on a large scale by filtering a mixture of alcohol and yeast through a cask filled with beech shavings soaked in vinegar. As the fermenting alcohol slowly trickles down, it comes in close contact with the air, absorbing O so rapidly that sometimes before it reaches the bottom it becomes entirely converted into vinegar.

Cider Vinegar.—Cider contains some nitrogenous matter, which acts as a ferment, by which the starch of the apple is broken up into C_2H_6O and CO_2. This makes what is called "*old cider.*" By exposure to the air and heat, the alcohol passes on to the second stage, and the acetic acid formed produces the sour taste of the vinegar.*

Properties.—Acetic acid is a solvent of albumen, gelatin, fibrin, etc. Hence it takes from meat, eggs, oysters, etc., their most strengthening constituents. For a similar reason, vinegar is a valuable assistant in digesting such food. It allays thirst, and was anciently carried by the Roman soldiers in a little flask for that purpose. Sugar added to vinegar quickly passes to the second stage of fermentation, and increases its strength. Indeed, vinegar is sometimes made entirely from sweetened water and tea-leaves, which act as a ferment. Vinegars of commerce are often sharpened by the addition of H_2SO_4 and pungent spices.†

Preserves frequently "work," as it is called, and then sour. The bubbles of gas which rise to the surface indicate the alcoholic fermentation. If neglected, this soon passes to the acetic stage. It may be checked by scalding, which destroys the ferment.

Aldehyde.—A molecule of alcohol absorbs two atoms of O from the air, as we have seen, forming acetic acid and water. In this process, there is an intermediate step during which the two atoms of H combine with one atom of O, forming a molecule of water. This intermediate

* Mother, in vinegar, is a fungus (*Mycoderma aceti*) produced by the decomposition of the nitrogenous matter. It absorbs O from the air and gives it up to the alcohol.

† We can easily detect these by evaporating a half-gill in a saucer, placed over hot water. As it boils down, add a little sugar, taking care not to allow it to burn. If the liquid turns black, it is proof of the presence of H_2SO_4. As the last evaporates, the odor of cayenne pepper, etc. (if there be any), can be readily distinguished.—In England, commercial vinegar is permitted by law to have one part in a thousand of oil of vitriol, as this keeps it from moulding.

substance is called *aldehyde*,* which quickly absorbs a second atom of O, producing acetic acid.

ORGANIC RADICALS.

A RADICAL is the root of a series of compounds which differ from each other by a constant amount. Such series are said to be homologous (*homos*, same; *logos*, proportion), and the different members to be homologues of each other.

Marsh-gas Series.—Marsh-gas, CH_4 (page 81), is the first member of a series of hydrocarbons, whose common difference is CH_2. The symbol CH_4 may be written $(CH_3)H$, and considered as the hydride of a group of atoms called *methyl:* hence marsh-gas is called *methyl hydride.* Methyl is the radical of a series of compounds, and plays the part of an element in various chemical reactions. The other members of the series may be written in the same way, and are regarded as hydrides of the radicals *ethyl, propyl, butyl*, etc.

Methyl Hydride (Marsh-gas).........	CH_4	$=CH_3,H$
Deutyl or Ethyl Hydride............	C_2H_6	$=C_2H_5,H$
Trityl or Propyl Hydride...........	C_3H_8	$=C_3H_7,H$
Tetryl or Butyl Hydride............	C_4H_{10}	$=C_4H_9,H$
Pentyl or Amyl Hydride............	C_5H_{12}	$=C_5H_{11},H$
Hexyl Hydride....................	C_6H_{14}	$=C_6H_{13},H$
Heptyl Hydride....................	C_7H_{16}	$=C_7H_{15},H$
Octyl Hydride.....................	C_8H_{18}	$=C_8H_{17},H$
Nonyl Hydride....................	C_9H_{20}	$=C_9H_{19},H$
etc.		etc.

* The odor of aldehyde may be obtained by holding a red-hot coil of Pt wire in a goblet containing a few drops of alcohol. This experiment, showing the

The Alcohols.—Common alcohol, C_2H_6O, may be written C_2H_5,HO, which is the hydrate of the radical ethyl; hence it is known as the *ethyl hydrate.* Each of the other radicals just named has its hydrate or alcohol. In some cases it has not yet been separated, though its symbol is known, and the body is believed to exist. The series has been continued as high as melyl alcohol, $C_{30}H_{62}O$.

			Boiling Point.
Methyl Alcohol (wood-spirit) *....	CH_4O	$=CH_3,HO$	.... 66° C
Ethyl Alcohol..................	C_2H_6O	$=C_2H_5,HO$	.... 78°
Propyl Alcohol.................	C_3H_8O	$=C_3H_7,HO$	.... 96°
Butyl Alcohol..................	$C_4H_{10}O$	$=C_4H_9,HO$	....109°
Amyl Alcohol (fusel oil) †........	$C_5H_{12}O$	$=C_5H_{11},HO$	....132°
Hexyl Alcohol..................	$C_6H_{14}O$	$=C_6H_{13},HO$	....150°
Heptyl Alcohol.................	$C_7H_{16}O$	$=C_7H_{15},HO$	....164°
———	$C_8H_{18}O$	$=C_8H_{17},HO$	....
———	$C_9H_{20}O$	$=C_9H_{19},HO$	....
Decatyl Alcohol................	$C_{10}H_{22}O$	$=C_{10}H_{21},HO$	....212°

Aldehydes and Acids.—Ethyl, or common alcohol, by an oxidizing agent loses two atoms of H, and is changed, as we have seen, to C_2H_4O, or ethyl aldehyde. This can be further oxidized into acetic acid. Each of

formation of aldehyde from alcohol, may be very profitably followed by another, illustrating the change of alcohol into acetic acid. Place a little platinum black in a watch crystal, near a small cup of alcohol. Cover them both with a glass receiver, and set them in the sunlight. Soon a mist will gather, and tiny streams of the condensed vapor of acetic acid will collect and run down the sides of the glass. Fresh air must be occasionally admitted to oxidize the alcohol.

* Methyl alcohol is very like ethyl alcohol, and is used for similar purposes, as dissolving resins, etc. *Methylated spirit* is ethyl alcohol with about 10 per cent. of wood spirit, which does not injure it for many uses to which ordinary alcohol is applied.

† This is formed in distilling whiskey from potatoes. It is present in common C_2H_6O, giving a slightly unpleasant odor when it evaporates from the hand. It is extremely poisonous, and as it is often contained in liquors, must greatly increase their destructive and intoxicating properties.

the alcohols can thus be oxidized, and will yield an aldehyde and an acid. The following is a list which is more complete even than that of the alcohols themselves:

		Source.
Formic Acid*..........	CH_2O_2	......Red ants, oxalic acid, etc.
Acetic Acid............	$C_2H_4O_2$	......Alcohol, distillation of wood.
Propionic Acid.........	$C_3H_6O_2$	......Glycerin.
Butyric Acid...........	$C_4H_8O_2$	......Butter.
Valerianic Acid.........	$C_5H_{10}O_2$	......Valerian root.
Caproic Acid...........	$C_6H_{12}O_2$	......Butter.
Œnanthylic Acid........	$C_7H_{14}O_2$	......Castor oil.
Capric Acid............	$C_8H_{16}O_2$	......Butter.
Pelargonic Acid........	$C_9H_{18}O_2$	......Geranium leaves.
Rutic Acid.............	$C_{10}H_{20}O_2$	......Oil of rue, butter.
Lauric Acid............	$C_{12}H_{24}O_2$	......Berries of bay tree.
Myristic Acid..........	$C_{14}H_{28}O_2$	......Nutmeg butter.
Palmitic Acid†.........	$C_{16}H_{32}O_2$	......Palm oil.
Margaric Acid†.........	$C_{17}H_{34}O_2$	......Animal fats.
Stearic Acid†..........	$C_{18}H_{36}O_2$	...... " "
Arachic Acid...........	$C_{20}H_{40}O_2$	......Butter.
Behenic Acid...........	$C_{22}H_{44}O_2$	......
Hyænic Acid............	$C_{25}H_{50}O_2$	......
Cerotic Acid...........	$C_{27}H_{54}O_2$	......Beeswax.
Melissic Acid..........	$C_{30}H_{60}O_2$	...... "

The Ethers.—Common ether is formed, as we have seen, from ethyl alcohol by the action of H_2SO_4; and its symbol, $(C_2H_5)_2O$, represents that it is considered as *ethyl oxide.* Each of the other alcohols ‡ has its ether—the oxide of its radical. Thus,

* Formic acid, which was found in red ants (*Formica rufa*), is a fiery, pungent fluid, which blisters the skin. It is made from methyl alcohol, as acetic acid is from common or ethyl alcohol.

† Palmitic, margaric, and stearic acids, are known as the fatty acids (see page 218), and are of great value in the arts.

‡ The term *alcohol* is now applied to those neutral compounds of H, C, and O, which react upon the acids, eliminating water and forming ethers.

Methyl Ether.................$C_2H_6O = (CH_3)_2O$
Ethyl Ether..................$C_4H_{10}O = (C_2H_5)_2O$
Propyl Ether.................$C_6H_{14}O = (C_3H_7)_2O$
Butyl Ether..................$C_8H_{18}O = (C_4H_9)_2O$
Amyl Ether...................$C_{10}H_{22}O = (C_5H_{11})_2O$

Compound Ammonias.—Each alcohol also forms a series of compound ammonias. H_3N may be written thus, $\left.\begin{matrix}H\\H\\H\end{matrix}\right\}N$, and one or more atoms of H may be replaced by a radical. In the ethyl series, for example, we have $\left.\begin{matrix}C_2H_5\\H\\H\end{matrix}\right\}N$, or ethylamine; $\left.\begin{matrix}C_2H_5\\C_2H_5\\H\end{matrix}\right\}N$, diethylamine; $\left.\begin{matrix}C_2H_5\\C_2H_5\\C_2H_5\end{matrix}\right\}N$, triethylamine. These ammonias closely resemble common ammonia, neutralize acids, produce white clouds with HCl, and form crystallizable salts; though they steadily rise in their boiling point.

Salts of the Radicals.—The symbol for water may be written thus, $\left.\begin{matrix}H\\H\end{matrix}\right\}O$; caustic potash, $\left.\begin{matrix}K\\H\end{matrix}\right\}O$; and ethyl alcohol after the same type, $\left.\begin{matrix}C_2H_5\\H\end{matrix}\right\}O$. As by adding HCl to KHO we obtain KCl and H_2O, so we can form the chlorides, iodides, bromides, etc., of all the radicals of the marsh-gas series by treating the alcohol with the proper acid. Ethyl, for example, thus furnishes compounds analogous to the potassium salts.

1. Potassium nitrate.........................$\left.\begin{matrix}K\\NO_2\end{matrix}\right\}O.$
1. Ethyl nitrate (Ethyl-nitric ether)..........$\left.\begin{matrix}C_2H_5\\NO_2\end{matrix}\right\}O.$
2. Hydrogen-potassium-sulphate.................$\left.\begin{matrix}K\\H\end{matrix}\right\}SO_4.$
2. Hydrogen-ethyl-sulphate (Sulphuric ether*).$\left.\begin{matrix}C_2H_5\\H\end{matrix}\right\}SO_4.$

* This, rather than ethyl oxide (see p. 197), is the true sulphuric ether; com-

3. Potassium sulphate........................ $\left.\begin{matrix}K\\K\end{matrix}\right\} SO_4$.

3. Ethyl sulphate (Ethyl-sulphuric ether)..... $\left.\begin{matrix}C_2H_5\\C_2H_5\end{matrix}\right\} SO_4$.

4. Potassium acetate....................... $\left.\begin{matrix}C_2H_3O\\K\end{matrix}\right\} O$.

4. Ethyl acetate (Ethyl-acetic ether)......... $\left.\begin{matrix}C_2H_3O\\C_2H_5\end{matrix}\right\} O$.

The salts of the radicals are often termed *compound ethers,* to distinguish them from the simple ethers, which are the *oxides.* They are extensively sold as flavoring extracts for the use of confectioners and cooks. The essence of jargonelle pear is an alcoholic solution of amyl acetate; apple oil, of amyl valerianate; pine apple, of ethyl butyrate.

There are many series of organic bodies, of which those given are merely illustrations. The various changes which they can undergo indicate what a wide field lies open for discovery, and the multitude of possible organic compounds.

DESTRUCTIVE DISTILLATION.

1. DECAY.—When wood decays slowly in the open air, the H passes off, the proportion of C increases, the color darkens, and a black carbonaceous mass remains, called *humus.* This is of great value to the soil, as its pores absorb H_3N, which, together with CO_2 produced by

pare its formula with that of sulphuric acid, showing it to be a true salt as defined in the note on page 23.

its decay, is furnished to the growing plant. When the supply of humus is exhausted from the soil, we restore it by adding straw, etc., and by plowing in green crops.

2. DISTILLATION.—When hard wood, as beech or oak, is heated to a high temperature, with no O present, or an imperfect supply, it is decomposed; the charcoal remains, while a large number of products is formed, among which are H, CO, CO_2, H_2O, CH_4, methyl alcohol, acetic or pyroligneous acid, crėosote, paraffine, tar, etc.

Pyroligneous Acid (wood-vinegar) is the crude acetic acid. It is used in making the acetates from which the pure acid is obtained by the action of a stronger acid, as H_2SO_4.

Creosote (flesh-preserver) is a colorless, poisonous liquid, with a flavor of burnt wood.* It has powerful antiseptic properties. It imparts to smoke a characteristic odor, renders it irritating to the eyes, and also gives to it the power which it possesses of curing hams, beef, etc.

Paraffine (*parum*, little; *affinis*, affinity), so called because the acids and bases have no effect upon it, is a hard, white, tasteless solid, resembling spermaceti. It forms beautiful candles, which look and burn like the finest of wax. It is a product of the rectification of beech-wood tar, but the paraffine of commerce is now obtained from petroleum.

Tar is made, like charcoal, by burning heaps of wood under a covering of earth which excludes the air: an imperfect combustion ensues, the resinous matter exudes,

* **Much of that which is sold as creosote is carbolic acid.** **(See page 206.)**

and, trickling down, runs into a reservoir below. On the extensive pine-barrens of North Carolina the tar of commerce is principally produced.

COAL-TAR is formed in the process of making coal-gas. (See p. 81.) It was formerly thought valueless, but is now used for a great variety of purposes. On distillation it yields, among other products, ammonium salts (see pp. 83, 135), carbolic acid, benzole, coal-naphtha, and dead-oil.

Carbolic Acid (*Phenic Acid,** C_6H_6O) is noted for its antiseptic and disinfecting properties. By heating it with HNO_3, *picric acid* is formed. This colors a rich yellow, and is a very popular silk dye. The picrates are yellow, explosive salts. Potassium picrate is used in making certain kinds of gunpowder.

Benzole † is a light oil used as a solvent of gutta-percha, caoutchouc, and wax; for removing grease-spots; as a burning-fluid, etc. Its ready inflammability makes it an exceedingly dangerous article. A current of air passed through benzole, as well as through any of the other light hydrocarbons, will absorb so much vapor that it may be burned as an illuminating gas. Machines for lighting houses, etc., are based upon this principle.

Nitro-Benzole is formed by treating benzole with HNO_3. It is a heavy, oily liquid, with an odor like that of bitter almonds. It is chiefly valuable, however, as the source of aniline,‡ from which are prepared the celebrated

* The formula for carbolic acid is C_6H_6O, and may be written as C_6H_5,HO. The acid may then be considered as the hydrate of the radical phenyl, and hence is sometimes called *phenyl alcohol*. Phenyl is the radical of a series of hydrocarbons, abundant in the coal-oils.

† Benzole was so named because of its abundance in benzoic acid. It was formerly sold as benzine, but a cheaper coal-oil has now taken its place.

‡ In 1856, Mr. Perkin, while experimenting with aniline in hopes of making quinine, treated it with potassium bichromate. He did not succeed in his

coal-tar dyes.—*Example:* mauve, magenta, etc. Who but a chemist would have searched in black, sticky coal-tar for these rainbow-tints, the stored-up sunshine of the carboniferous age!

Naphtha is a volatile, limpid oil, with a peculiar odor and generally a light straw color. It is composed of several hydrocarbons and is very inflammable. *Naphthaline*

attempt, but he obtained a beautiful purple dye, which was soon introduced to commerce under the name of *mauve*. A host of imitators at once sought to obtain the color without using potassium bichromate. As the only use of the latter was to oxidize the aniline, they reasoned that they might use any other oxidizing agent. Arsenic, among other substances, was tried, but instead of a purple the red known as *magenta* was the result. The coloring matter, however, does not contain any arsenic; being a salt of a base called *rosaniline*. Rosaniline itself is colorless, and reveals its magnificent tints only in its compounds. "The crystals of its salts exhibit by reflected light the metallic green color of beetles' wings, but are of a deep red color when seen by transmitted light." Magenta is manufactured on an enormous scale in England, more as a substance from which to obtain other dyes than for direct use in dyeing. A single firm produces twelve tons a week. The quantity of magenta furnished by one hundred pounds of coal, is very small; but this is compensated for by its intense coloring power, since it will dye a quantity of wool nearly equal in weight to the coal. In making magenta on the large scale there are considerable quantities of residual products. These of course have been examined with a view to further profit, and the result has been the discovery of a beautiful orange color called *phosphine*. This is much used to produce scarlet, by first dyeing the silk or wool with magenta, and then passing it through a bath of phosphine. By treating magenta with aniline, a beautiful blue is obtained. This is insoluble in water, but is rendered soluble exactly as indigo is, by treating it with sulphuric acid. Another curious dye formed from aniline is known as *Nicholson's blue*. This is completely decolorized by alkalies, and the color is restored by acids. In dyeing with it, the silk or wool is first immersed in a colorless solution of the dye, and then dipped into dilute sulphuric acid, when the blue is at once developed. If magenta is heated with iodide of ethyl or methyl, an excess of the iodide being employed, a most beautiful green is the result. If, however, this green is heated sufficiently to drive off the excess of iodide, a violet color is the result; so that it will not do for ladies wearing dresses dyed with this green to sit too near the fire. After all the coloring matter has been extracted from the aniline, a residue remains which has an intense black color and is largely used for making printing ink. Very few of the aniline colors when in powder give a person any idea of the color which they will produce when moistened. Magenta, for instance, when dry, is a beautiful green with a bronze-like lustre. It is a pretty experiment to coat a sheet of glass with one of these colors, which is readily done by dissolving in alcohol (Hofmann's violet being the best) and allowing a film of it to evaporate on the glass. When seen by transmitted light it is of a beautiful violet, but with reflected light it displays a tint rivalling in brilliancy the tail of a peacock.—*Boston Journal of Chemistry.*

is a crystalline solid occurring in beautiful pearly scales. It is especially abundant in dead-oil, and may be formed by passing olefiant gas or benzole through red-hot tubes. *Anthracene* accompanies naphthaline in the latter part of its distillation. It is also a white solid. It is of interest since the coloring principle of madder—alizarine—has been made from it.

Dead-Oil is used for preserving timber; as a cement for roofs and walls; for oiling machinery, etc.

PETROLEUM (*petra,* a rock; *oleum,* oil) is probably the product of the distillation of organic matter beneath the surface of the earth. It is not always connected with coal, as it is often found outside the coal-measures, as in New York and Canada. The distillation must have taken place at a much greater depth than that at which the oil is now found, as it would naturally rise through the fissures of the rock and gather in the cavities above. Sometimes the oil has collected on the surface of subterranean pools of salt-water, so that after a time the oil is exhausted, and salt-water only is pumped up; or if the well strikes the lower part of the cavity, the water will first be pumped and afterward the oil. The crude oil from the well is purified by distillation. That which passes over at the lowest temperature is called *naphtha:* as the heat is increased, there passes off next *kerosene oil** for illumination, and lastly *lubricating oil.* The

* Kerosene accidents generally rise from the presence of naphtha. This is a cheap, light, dangerous oil. Its vapor, however, is not explosive unless mixed with air. While a lamp, which contains adulterated kerosene, is burning quietly there is no danger. The vapor rises from the oil, fills the empty space in the lamp, but being unmixed with air cannot explode. Let, however, a draught of cold air strike it, or carry it into a cold room—instantly the vapor will be condensed, the air will rush in, and a dangerous mixture be formed. Or when the light is extinguished at night the vapor will cool, air pass in, and a mixture be

kerosene is deodorized and decolorized by the use of H_2SO_4 and other chemicals, which are stirred in the oil, after which it is redistilled.

Bitumen or Asphaltum.—Petroleum and naphtha, flowing from the ground, have formed beds of bitumen in various parts of the world. This change is caused by a gradual oxidation and hardening, as turpentine changes to rosin. Tar Lake is situated on the island of Trinidad. It is nearly three miles in circumference. The bitumen is used for the same purposes as pitch, which it closely resembles. Near the shore it is hard and compact, except in hot weather, when it becomes sticky. At the centre it is soft, and fresh bitumen boils up to the surface. Asphaltum is found in immense quantities in California and in Canada. It is a natural cement for laying stone or brick. It was used in building the walls of Babylon, for which purpose it was gathered from the fountain of Is on the banks of the Euphrates. It was a prominent ingredient in the "Greek Fire," so much used by the nations of Eastern Europe in their naval wars, even as late as the fourteenth century. This consisted of bitumen, sulphur, and pitch, and was thrown through long, copper tubes, from hideous figures erected on the prow of the vessel. Bitumen is used in making the famous promenades of the Boulevards in Paris.

produced which will be ready to explode when the lamp is relighted. Kerosene of the legal standard is no more explosive than water, and will even extinguish a flame applied to it at the ordinary temperature. Dr. Nichols gives the following simple test: *Fill a bowl partly full of hot water. Insert a thermometer, and add cold water until the temperature is 110° F. Then pour into the bowl a spoonful of kerosene and apply a lighted match. If it takes fire the oil contains naphtha and is dangerous; if not, the kerosene may be used with perfect safety.*

THE ORGANIC ACIDS.

THERE are many vegetable acids found native in plants —generally, however, combined with some base.

Oxalic Acid ($C_2H_2O_4$) is familiar in the sour taste of rhubarb, sorrel, etc. In these plants the acid is combined with K and Ca. It may be prepared by the action of HNO_3 on sugar.* It is a potent poison. The antidote is powdered magnesia, or chalk, stirred in H_2O. It is a test of lime, forming a delicate white precipitate of calcium oxalate. A solution of oxalic acid is much used to remove ink stains, and is often sold for this purpose under the deceptive name of "salts of lemon." The acid unites with the Fe of the ink, and the iron oxalate thus made is soluble in H_2O. It should be washed out immediately, as it will corrode the cloth.

Tartaric Acid ($C_4H_6O_6$) exists in many fruits, principally in the grape, combined with K as hydrogen potassium tartrate ("bitartrate of potash"). This settles during the fermentation of wine (see p. 195), and when purified is called *cream of tartar,* from which tartaric acid is made. It forms transparent crystals of a pleasant acid taste, which are permanent in the air. Its aqueous solution gradually becomes mouldy and turns into acetic acid. *Tartar emetic* is an antimony potassium tartrate. *Rochelle salt* is a sodium potassium tartrate; it is commonly used in medicine in the form of *Seidlitz powders.* These are contained in a blue and a white paper. The former

* Oxalic acid is made on a large scale from sawdust, soda, and caustic potash. The woody fibre is resolved into oxalic acid, which combines with the bases, forming sodium and potassium oxalates. From these the acid is readily obtained. Sawdust will yield more than half its weight of crystals of this salt.

holds 120 grains of Rochelle salt, and 40 grains of bicarbonate of soda; the latter 35 grains of tartaric acid. They are dissolved in separate goblets. The one containing the acid is emptied into the other, when the CO_2 is set free, producing a violent effervescence and disguising the taste of the medicine.

Malic Acid ($C_4H_6O_5$) occurs abundantly in most acid fruits, particularly in unripe apples, whence its name from *malum*, an apple. *Citric acid* (*citrus*, a lemon), the acid of the lemon, lime, etc., is often found associated with it, as in the gooseberry, raspberry, and strawberry. Citric acid is used in medicine as magnesium citrate.

Tannic Acid [*tannin* ($C_{27}H_{22}O_{17}$)] is found in the leaf and bark of trees.*—*Example:* Oak, hemlock. Nut-galls are excrescences which are formed by the puncture of an insect on the leaves of a certain species of oak. Tannin has an astringent taste, is soluble in water, and hardens albuminous substances, as gelatine.

Tanning.— After the hair has been removed from the skins by milk of lime, they are soaked for days, the best kinds for months, in vats full of water and ground oak or hemlock bark (tan-bark). The tannic acid of the bark is dissolved, and entering the pores of the skin, unites with the gelatine, forming a hard, insoluble compound which is the basis of leather. Leather is blackened by washing the hide on one side with a solution of copperas. The tannic acid unites with the iron, forming a tannate of iron—an ink. In the same way, drops of tea on a knife-blade stain it black.

* This astringent principle is widely diffused. There are several compounds which possess similar properties, yet differ in chemical composition. The tannin of the oak is called *quercitannic acid;* that of nut-galls, *gallotannic acid;* that of tea, *theitannic*, and that of coffee, *caffeotannic acid*.

Ink is made by adding a solution of nut-galls to one of copperas. The iron tannate thus formed has a pale blue-black color, as in the best writing-fluids; by exposure to the air, the Fe absorbs more O, the ink darkening in color until it is a deep black. Gum is added to thicken and regulate the flow of the fluid from the pen. Creosote or corrosive sublimate is used to prevent mouldiness. Steel pens are corroded by the free H_2SO_4 contained in the ink, but gold pens are not affected by it.*

Gallic Acid ($C_7H_6O_5$) is associated with tannin in nut-galls, and can be formed from tannic acid. *Pyrogallic acid* can be obtained by the sublimation of gallic or gallotannic acid. It is extensively used in photography for the purpose of developing the latent image in the collodion film after exposure to the action of the light. (See p. 167.)

THE ORGANIC BASES.

THE organic bases, or alkaloids, as they are called, are the bases of true salts found in plants. They dissolve slightly in H_2O, but freely in alcohol. They have a bitter taste, and rank among the most fearful poisons and valuable medicines. All the alkaloids contain N.†

* The following is an instructive experiment, illustrating the manner of making ink, of removing stains with oxalic acid, and also the relative strength of the acids and alkalies. Take a large test-tube, and add the following reagents in solution, *cautiously, drop by drop*, watching the result and explaining the reactions: 1, iron sulphate (*copperas*); 2, tannic acid (*tannin*); 3, oxalic acid; 4, sodium carbonate (*sal-soda*); 5, hydrochloric acid (*muriatic*); 6, ammonia (*hartshorn*); 7, nitric acid (*aquafortis*); 8, caustic potash; 9, sulphuric acid (*oil of vitriol*).

† A convenient antidote is tannin, which forms with nearly all of them insoluble curdy tannates. Almost any liquid containing it is of value—as strong,

Opium is the dried juice of the poppy plant, which is extensively cultivated in Turkey for the sake of this product. Workmen pass along the rows soon after the flowers have fallen off, cutting slightly each capsule. From these incisions a milky juice exudes and collects in little tears. These are gathered and wrapped in leaves for the market. Opium contains six different alkaloids in combination with a single acid. In small doses, opium is a sedative medicine; in larger ones, a narcotic poison. *Laudanum* is the tincture of opium; and *paregoric,* a camphorated tincture flavored with aromatics.

Opium-eating. — Opium produces a powerful influence on the nervous system. It stimulates the brain and excites the imagination to a wonderful pitch of intensity. The dreams of the opium-eater are said to be vivid and fantastic beyond description. The dose must, however, be gradually increased to repeat the effect, and the result is most disastrous. The nervous system becomes deranged, and no relief can be secured save by a fresh resort to this baneful drug.* Labor becomes irksome,

green tea. This is also of use as it tends to keep the patient awake, the great necessity in the case of a narcotic poison.

* In time, the whole system becomes so impregnated with it that additional doses utterly fail to produce the delightful effect which at first so fascinated the victim. Then, while combining with the nerves, it set free a vast amount of vitality and force, but now it has satisfied itself. The subtle alkaloid has worked its way into the tissues and coatings of his entire internal organism. If, resolutely, he summons his enfeebled will, and commences the conflict, an agony of endurance, which defies all description, is before him. The whole body must be reorganized, and, atom by atom, the life-energy of the man must drag out of the flesh and blood the fearful poison. If, too weak to attempt so terrible a struggle, he continues the use of the fatal drug, he moves on directly to one fate, the opium-eater's grave. Paregoric, laudanum, morphine, and the different preparations of opium are in almost every case taken first as a sedative from pain or fatiguing labor, with no thought of becoming addicted to their use. But so insinuating is it that the victim forms the habit ere he is aware, and only knows he is a slave when for some reason he attempts to cease the customary dose. No person can be too careful in the use of a narcotic whose influence is liable to become so destructive.

ordinary food distasteful, and racking pains torment the body.

Morphine (*Morpheus,* the god of sleep) is one of the alkaloid bases of opium, and like it is used to alleviate pain and produce sleep. It is usually given as a sulphate or chloride.

Quinine is prepared from Peruvian bark. A tincture of the bark, or sulphate of quinia, is employed in medicine in cases of fever and ague and other periodic diseases, and also as a tonic.

Nicotine is the active principle of the tobacco plant, of which it forms from 2 to 8 per cent. It is volatile, and passes off in the smoke. A drop will kill a large dog. It probably produces many of the ill effects which follow the use of tobacco.

Strychnine is prepared from the nux vomica and the St. Ignatius's bean. It is also a constituent of the celebrated upas poison.* "It is so intensely bitter that one grain will impart a flavor to twenty-five gallons of water. One-thirtieth of a grain has killed a dog in thirty seconds, while half a grain is fatal to man."

The *Chromatic Test,* as it is called, consists in placing on a clean porcelain plate a drop of the suspected liquid, a drop of H_2SO_4, and a crystal of potassium bichromate. Mix the three very slowly with a clean glass rod. If there be any strychnine present, it will change the color into a beautiful violet tint, passing into a pale rose.†

* "The 'woorara,' with which the South American Indians poison their arrows, is a variety of strychnine. This is so deadly that the scratch of a needle dipped in it will produce death; yet it may be swallowed with impunity."—MILLER.

† Arsenic was once in favor with the poisoner, but Marsh's test infallibly re-

Caffeine and Theine constitute the active principle of tea * and coffee, and are isomeric. They crystallize in long, flexible, silky needles. In addition, tea contains from 13 to 18 per cent. of a form of tannin (see p. 211, note), about 15 per cent. of a nitrogenous substance allied to caseine,† and a volatile oil which gives to it its aromatic odor and taste. Coffee contains about 14 per cent. of a fixed oil, and also an essential oil which is developed in roasting, and is very volatile, so that it will soon escape unless the coffee be kept tightly covered.

veals its presence in the body of the victim, even after many years have elapsed. The organic poisons are so easily acted upon by the fluids of the system, that in one case, though four grains were taken, and death ensued very quickly, yet the "chromatic test" failed to detect the presence of strychnine in the stomach. However, the murderer is not to escape. This is the only poison except brucite (and that also is extracted from nux vomica) that produces tetanus or lock-jaw. This symptom proves to the physician that death has been caused by this alkaloid. To exhibit the effect of the poison a frog may be brought into the court-room and made to show its action. So sensitive is this little animal, that a few drops of oil containing only a hundred-thousandth of a grain of strychnine will instantly throw it into a rigid locked-jaw, in which it is incapable of the least motion.

* *Tea-raising.*—Tea-plants resemble in some respects the low whortleberry bush. They are raised in rows, three to five in a hill, very much as corn is with us, but they are not allowed to grow over one and a half feet high. The medium-sized leaves are picked by hand, the largest ones being left to favor the growth of the bushes. Each little hill or clump will furnish from three to five ounces of green leaves, or about one ounce of tea, in the course of the season. The leaves are first wilted in the sun, then trodden in baskets by barefooted men to break the stems, next rolled by the hands into a spiral shape, then left in a heap to heat again, and finally dried for the market. This constitutes *Black Tea*, and the color would be produced in any leaves left thus to wilt and heat in heaps in the open air. The Chinese always drink this kind of tea. They use no milk or sugar, and prepare it, not by steeping, but by pouring hot water on the tea and allowing it to stand for a few minutes. Whenever a friend calls on a Chinaman, common politeness requires that a cup of tea be immediately offered him.

Green Tea is prepared like black, except that it is not allowed to wilt or heat, and is quickly dried over a fire. It is also very frequently, if not always, colored —cheap black teas and leaves of other plants being added in large quantities. In this country, damaged teas and the "grounds" left at hotels are re-rolled, highly colored, packed in old tea-chests, and sent out as new teas. Certain varieties of black tea even receive a coating of black-lead to make them shiny.

† This is lost in the "grounds." The Japanese, however, eat the tea-leaves, and so save this nutritive part.

ORGANIC COLORING PRINCIPLES.

THE organic coloring principles are generally of vegetable origin. They are found in the roots, wood, bark, flowers, and seeds of plants.

Dyeing.—Very few of the colors have such an affinity for the fibres of the cloth that they will not wash out. Those which, like indigo, will dye directly, are called *substantive* colors. But the majority are *adjective* colors which require a third substance having an attraction for both the coloring matter and the cloth, to hold them together. Such substances are called *mordants* (*mordeo,* to bite), because they bind the dye in the cloth, thus making a "fast color." The most common mordants are alum, tin oxide, and copperas. In dyeing, the cloth is first dipped into a solution of the mordant, and then into one of the dye-stuff. The mordant, by means of a stamp, may be applied to the cloth in the form of a pattern, and when it is afterward washed, the color will be removed except where the mordant fixed it in the printed figure. The same dye will produce different colors by a change of mordants.—*Example:* Madder with iron gives a fine purple, with alum a pink, and with iron and alum a chocolate. This principle lies at the basis of dyeing "prints." *

* A calico-printing machine is very complex. The cloth passes between a series of rollers, upon which the corresponding mordant is put, as ink is on type. A single machine sometimes prints from twenty sets of rollers; yet each impression follows the other so accurately, that when the cloth has passed through, the entire pattern is printed upon it with the different mordants more perfectly than any painter could do it, and so rapidly that a mile of cloth has been printed with four mordants in an hour. The cloth when it leaves the printing machine, though stamped with the mordants in the form of the figure, betrays nothing of

Coloring Substances.—*Madder* is the root of a plant found in the East, and extensively cultivated elsewhere. When first dug it is yellow, but by exposure it absorbs O and becomes red. It is used in dyeing the brilliant Turkey-red. The coloring principle, which is named *alizarine,* is said to be identical with that derived from anthracene, a hydrocarbon found in coal-tar (see p. 208). *Cochineal* is a dried insect that in life lives upon a species of cactus in Central America. The coloring matter is called *Carmine.* It yields the brightest crimson and purple dyes.* *Brazil-wood* furnishes a red which is not very permanent. It is used for making red ink. The *indigo* of commerce is obtained from a bushy plant found in the East Indies. By fermenting for some days in vats of water, the coloring matter is developed. Reducing agents change indigo into a soluble and colorless substance by the absorption of H.† It is then called "white indigo." In this form it is extensively used in dyeing. The cloth becomes permanently colored on exposure to the air, when the insoluble blue indigo is formed in its fibres. *Logwood* is so named because imported in logs. It is the heart of a South and Central American tree. With a mordant of alum it dyes black. *Litmus* is obtained from a variety of lichens common along the southern coast of

the real design until after being dipped in the dye, which acting on the different mordants brings out the desired colors. The print is now washed, glazed, and fitted for the market.

* The purple of which we read in ancient writings was a secret with the Tyrians. King Huram, we learn, sent a workman to Solomon skilled in this art. The dye was obtained from a shell-fish that was found on the coast of Phœnicia. Each animal yielded a tiny drop of the precious liquid. "A yard of cloth dipped twice in this costly dye was worth $150."

† Dissolve a little indigo in strong H_2SO_4. Color a test-tube of H_2O with the solution. Add a drop of HNO_3, and on gently heating, the color will disappear.

Europe. Its juice is colorless, but on the addition of H_3N it assumes a rich purple blue. *Leaf-green* (chlorophyll), as found in plants, is a resinous substance containing several coloring matters. It seems to be developed by the action of the sunbeam. Plants removed from a dark cellar to the sunlight rapidly turn green.

THE OILS AND FATS.

THE oils and fats are derived from both the vegetable and the animal kingdom. They are divided into two classes—*fixed* and *volatile*. The former make soaps, the latter do not. The former, when heated above 500°, give off acrid and offensive vapors;* the latter may be distilled without alteration.†

1. THE FIXED OILS.

Composition.—The fatty bodies are salts, being composed of *stearin*, *palmitin*, and *olein*.‡ These consist of three acids, stearic, palmitic, and oleic, combined with a common base—*glycerin*.

The first two of these salts are solids at common temperatures, and form fats; the last is a liquid, and forms oils. The relative proportion of olein contained

* At a higher temperature they are decomposed, and among the products is an acrid substance (*acrolein*) with which we are familiar in the disagreeable smell of a smouldering candle-wick and in burning fat.

† "The former produce a permanent stain on paper, the latter do not. A cork twisted into the neck of a bottle containing a fixed oil makes no noise; in a volatile oil it squeaks."

‡ Stearin, from *stear*, suet; palmitin, since it is abundant in palm oil; olein, from *oleum*, oil; glycerin, from *glukeros*, sweet.

in any fatty substance determines its fluidity.—*Example:* Stearin is abundant in tallow, and palmitin in butter; hence their comparative consistency. Lard, on the other hand, contains so much olein that it is expressed as "lard-oil." Olive-oil contains much olein and palmitin; the former remains fluid at ordinary temperatures, but the latter, in cold weather, hardens into a thick deposit, and renders the oil viscid.

Glycerin ($C_3H_8O_3$) is an odorless, transparent syrup. It is soluble in H_2O and alcohol. On account of its healing properties its use is common in dressing wounds, insect bites, chapped hands, etc.

By the action of HNO_3 and H_2SO_4 glycerin is converted into *nitro-glycerin* $[C_3H_5(NO_2)_3O_3]$, an oil that often explodes with fearful violence by a slight concussion. *Dynamite,* used in blasting, is powdered silex, or infusorial earth (*Geology,* p. 48), saturated with glycerin.

Soap.—If sweet-oil and H_2O be placed in a test-tube and shaken, they will mix but not unite; for on standing, the oil will rise to the top. Add, however, caustic potash or a little "lye" (see p. 128), when, on heating, a clear, soapy solution will be formed. The K of the alkali has combined with the oleic and palmitic acids of the oil, making two new salts—potassium oleate and potassium palmitate; while the expelled glycerin remains floating in the liquid.

The manufacture of soap is based on this principle. A variation in the alkaline base and the fat or oil used, produces the different kinds of soap. Potash, on account of its affinity for H_2O, forms *soft-soap.* Soda* is not deli-

* A deliquescent substance is one which dissolves in H_2O, which it absorbs from the air.

quescent, and hence makes *hard-soap*.* Lard forms a softer soap than tallow. *Castile soap* is made from olive oil and soda. Its mottled appearance is due to the coloring matter which is stirred through it while it is yet soft. *Home-made soap* is prepared by boiling "lye" and "grease."† As the latter contains such a variety of fatty substances the soap generally consists of the three salts—potassium oleate, palmitate, and stearate. *Yellow soap* contains some resin in place of fat. *Cocoa-nut-oil* makes a soap which will dissolve in salt water, as it contains an excess of alkali. *Soap-balls* are made by dissolving soap in a very little water, and then working it with starch to a proper consistency to be shaped into balls. *White toilet-soaps* are made from lard and soda. The *curdling of soap* in hard water is caused by the formation of a calcium or a magnesium soap which is insoluble in H_2O, and floats on the top as a greasy scum.‡

The Cleansing Qualities of Soap.—There exudes constantly from the pores of the skin an oily perspiration, and this catching the floating dust dries into a film which will not dissolve in H_2O. The alkali of the soap combines with this oily substance and makes a soluble soap. In addition, the alkali also dissolves the cuticle of the skin, and thus produces the "soapy feeling," as we term it, when we handle soap.

* Soap is frequently adulterated with gypsum, lime, pipe-clay, or sodium silicate. These may be detected by dissolving a piece of the soap in water or alcohol, and noticing if there be any precipitate.

† The heat hastens the chemical change, which takes place more slowly in making what is known as "cold soap."

‡ A soap made from lard, in water containing calcium carbonate, would undergo the following reaction: Potassium oleate + calcium carbonate = calcium oleate + potassium carbonate.

Saponification (*sapo*, soap; *facere*, to make) is the process of separating the fatty acids and glycerin, and is so named even when no soap is formed. One method is as follows: Tallow or lard is boiled with lime, and thus made into a calcium soap. This is decomposed by H_2SO_4, forming calcium sulphate, which, being insoluble, sinks to the bottom, leaving the three acids of the fat floating upon the surface.* The glycerin is also left by itself in the liquid, from whence it is removed and prepared for the market. The acids, when cool, are subjected to great pressure; the oleic flows out, leaving the stearic and palmitic acids as a milk-white, odorless, tasteless solid, which is commonly called *stearin*, and extensively used in the manufacture of *stearin* or *adamantine candles*.†

Wax is found in nearly all plants. It forms the shiny coating of the leaves and fruit.—*Example:* Lemon leaf, apple. Certain plants in Japan contain so much wax that it is separated by boiling and used for making candles. Bees, even when fed on sugar alone, have the power of converting it into wax, which is therefore to be regarded as an animal secretion.—MILLER. Beeswax is bleached by exposure to the air in thin ribbons.

Linseed Oil is a *drying oil*, as it is termed—*i. e.*, it

* Fat is also decomposed by the action of superheated steam, which at once liberates the fatty acids.

† *Paraffine candles* are made from coal-oil. *Wax candles* are manufactured by the following process: A large number of cotton wicks are hung upon a revolving frame with projecting arms. The wicks are fitted at the ends with metal tags to keep the wax from covering that part. As the machine slowly turns, a man, standing ready with a vessel of melted wax, carefully pours a little upon each wick in succession. This process is repeated until the candles reach the desired size. They are then rolled on a smooth stone slab, the tops cut by conical tubes, and the bottoms trimmed, when they are ready for use. The large tapers burned in Catholic cathedrals are made by placing the wick on a sheet of wax, rolling it up till the right thickness is reached, when the candle is trimmed and polished as before. *Spermaceti candles* are run from the white, crystalline, solid fat which is found with sperm oil in the head of the sperm whale.

absorbs O from the air,* and hardens by exposure. It is expressed from flaxseed, which furnishes about one-fifth of its weight in oil. *Boiled oil* is made by heating the crude oil with litharge, which entirely dissolves and greatly increases the drying property of the oil. Linseed oil is used in mixing paints and varnishes. *Putty* consists of linseed oil and whiting well mixed. *Printers' ink* is made by heating linseed oil until it becomes thick and viscid, when lampblack is added to give it the proper consistency.

Cod-liver Oil is extracted from the liver of the codfish. It contains I, Br, and P, and is much used as a remedy in consumption.

Croton Oil is made from the seeds of an Indian plant; it is a powerful medicinal agent.

Castor Oil is extracted from the castor-oil bean. It is used in medicine, and also in perfumery and hair-oils.

Sweet or Olive Oil is expressed from the olive fruit. It is an *unctuous oil, i. e.*, it absorbs O on exposure to the air—not hardening like the drying oils, but remaining sticky, and after a time becoming rancid.† In the southern part of Europe, olive-oil is extensively used as an article of food.

2. THE VOLATILE OILS.

THE volatile oils, unlike the fixed, make no soaps, and

* This absorption of O is sometimes so rapid as to be attended by sufficient heat to occasion the mass to take fire; and several serious conflagrations have been traced to such spontaneous combustion. (See p. 93.)

† This change is attended by a slight absorption of O, and appears to be due to the decomposition of certain mucilaginous and albuminous matters, which, during their decay, react on the fat, setting free the fatty acids, and decomposing the glycerin. Perfectly pure fats and oils do not become rancid.

dissolve readily in alcohol or ether. Their solution in alcohol forms an essence.

Sources.—The volatile or essential oils are of vegetable origin. They are found in the petals of a flower, as the violet; in the seed, as caraway; in the leaves, as mint, or in the root, as sassafras. Sometimes several kinds of oil are obtained from different parts of the same plant.—*Example:* In the orange tree, the flower, leaves, and rind of the fruit furnish each its own variety. The perfume of flowers is produced by these volatile oils; but how slight a quantity is present may be inferred from the statement that "one hundred pounds of fresh roses will give scarcely a quarter of an ounce of Attar of Roses."

Preparation.—In the peppermint, the wintergreen, and many others, the plant is distilled with water. The oils pass over with the steam, and are condensed in a refrigerator connected with the "Mint Still." The oil floats on the surface of the condensed water, and may be removed. A small portion, however, remains mingled with the latter, which thus acquires its peculiar taste and odor, constituting what is termed a "perfumed water." In some flowers, as the violet, jasmine, etc., the perfume is too delicate to be collected in this manner. They are therefore laid between woollen cloths saturated with some fixed oil. This absorbs the essential oil, which is then dissolved by alcohol. The oil of lemon or orange is obtained from the rind of the fruit by expression or by digesting in alcohol.

Composition.—$C_{10}H_{16}$ is the common symbol of a large number of these oils. Thus the oils of lemon, cloves, juniper, birch, black pepper, ginger, bergamot, turpentine, cubebs, oranges, etc., nearly twenty in all, are isomeric. They are pure hydrocarbons. A second class contains O, and a third S.

First Class of Volatile Oils.—The oil of turpentine ($C_{10}H_{16}$) is a type of this division. It is made by distilling pitch with H_2O. It is generally called *spirits of turpentine.* It is highly inflammable, and, owing to the excess of C, burns with a great smoke. By the union of two atoms of its H with an atom of the O of the air to form H_2O, it is converted into rosin.* *Camphene* is turpentine purified by repeated distillation. *Burning-fluid* is a mixture of camphene and alcohol. In the heat of the burning H of the latter, the C of the former is consumed, and this produces a bright light. The tendency of camphene to smoke is thus diminished, and the illuminating power increased. By the action of HCl on turpentine or oil of lemons an artificial camphor is produced which much resembles common camphor.

The Second Class includes camphor, the oils of bitter almonds, cinnamon, spearmint, wintergreen, etc. Camphor ($C_{10}H_{16}O$) is obtained by distilling the roots and leaves of the camphor-tree with water, and condensing the vapors on rice-straw. It is purified by sublimation. When kept in a bottle, it vaporizes, and its delicate crystals collect on the side toward the light. Taken internally, except in small doses, it is a virulent poison. Its solution in alcohol is called "spirits of camphor." If H_2O be added to this, the camphor will be precipitated as a flour-like powder.†

* In this way, the turpentine around the nozzle of a bottle in which it is kept becomes first sticky and then resinous. Old oil should not be taken to remove grease spots, as, while it will remove one, it will leave another of its own.

† Though camphor gum is powdered with difficulty, a few drops of alcohol will

The Third Class includes garlic, assafœtida, onions, mustard, horse-radish,* etc. They are known for their pungent taste and the disagreeable odor they often impart to the breath.†

THE RESINS AND BALSAMS.

THE resins are generally formed from the essential oils by a slow oxidation.—*Example:* Turpentine, as we have just seen, is changed to rosin, a resinous substance. If the resin is dissolved in some essential oil, it is called a balsam.—*Example:* Pitch is a true balsam, since by distillation it is separated into rosin and turpentine. They generally exude from incisions in trees and shrubs, in the form of a balsam, which oxidizes on exposure to the air, and becomes a resin.—*Example:* Spruce gum. The resins are translucent or transparent, brittle, insoluble in H_2O, but soluble in ether, alcohol, or any volatile oil, and form varnishes. They are non-conductors of electricity, and burn with much smoke. They do not decay, and, indeed, have the power of preserving other substances.‡

Rosin constitutes about 75 per cent. of *pitch,* a resinous substance which exudes from incisions made in the

remove all trouble. When small particles of powdered camphor are thrown on water free from grease, each fragment begins to dissolve with a remarkable gyratory motion, which is instantly checked by a drop of an essential oil allowed to fall upon the surface of the liquid.

* The essential oil of garlic, onions, etc., is the sulphide of allyl, a radical having the formula C_3H_5; the oil of horse-radish is the sulpho-cyanide of allyl.

† The oil of mustard is not contained in the seed, but is formed in it by the action of water and a latent ferment. This is the reason why mustard, when first prepared for the table, is bitter, but becomes pungent after a little time.

‡ For this reason they were used in embalming the bodies of the ancient Egyptians, which, after the lapse of two thousand years, are yet found dried into mummies in their mammoth tombs—the Pyramids.

trunks of certain species of pine. It is used in making soaps, to increase friction in violin-bows and the cords of clock-weights, and in soldering.

Lac exudes from the ficus-tree of the East Indies. An insect punctures the bark, and the juice flows out over the insect, which works it into cells in which to deposit its eggs. The dried gum incrusting the twigs is called *stick-lac;* when removed from the wood, *seed-lac;* when melted and strained, *shellac.* The liquefied resin is dropped upon large leaves, and so cools in broad, thin pieces. *Sealing-wax* is made of shellac and Venice turpentine; vermilion being added to give the red color. Shellac is much used in making varnishes.

Gum Benzoin also exudes from a tree in the East Indies. It is the principal source of benzoic acid. It is used in fumigation and in cosmetics, and on account of its fragrant odor is burnt as incense.*

Amber is a fossil resin which has exuded in some past age of the world's history from trees now extinct. It is sometimes found containing various insects perfectly preserved, which were without doubt entangled in the mass while it was yet soft. These are so beautifully embalmed in this transparent glass that they give us a good idea of the insect life of that age. Amber is cast up by the sea, principally along the shores of the Baltic; although it is also found in beds of lignite. It is commonly translucent, and susceptible of a high polish. It is used for ornaments, mouth-pieces, necklaces, buttons, etc.; and is an ingredient of carriage varnish.

* Place some green sprigs under a glass receiver, and at the bottom a hot iron, on which sprinkle a little benzoic acid. It will sublime and collect in beautifully delicate crystals on the green leaves above, making a perfect illustration of winter frost-work.

Caoutchouc or ***India-rubber*** (xC_5H_8) is a mixture of several hydrocarbons. It exudes from certain trees in South America as a milky juice.* The solvents of rubber are ether, naphtha, turpentine, chloroform, bisulphide of carbon, etc. It melts, but does not become solid on cooling. Freshly-cut surfaces readily cohere: this property, together with its power of resisting most reagents, renders it invaluable to the chemist in making flexible joints and tubes. "It loses its elastic power when stretched for a long time, but recovers it on being heated. In the manufacture of rubber goods for suspenders, etc., the rubber thread is drawn over bobbins and left for some days until it becomes inelastic. In this state it is woven, after which a hot wheel is rolled over the cloth to restore the elasticity."

Vulcanized Rubber is made by heating caoutchouc with a small amount of sulphur. This constituted Goodyear's original patent.† It is less liable to be hardened

* The globules of rubber are suspended in it as butter is in milk. By adding H_3N the sap may be kept unchanged for months, and is sometimes exported in that form preserved in tightly corked bottles. The tree, it is said, yields about a gill per day from each incision made. A little clay cup is placed underneath, from which the juice is collected and poured over clay or wooden patterns in successive layers as it dries. To hasten the process it is carried on over large open fires, the smoke of which gives to the rubber its black color; when pure it is almost white. When nearly hard, the rubber will receive any fanciful design which may be marked upon it with a pointed stick. The natives often form the clay into odd shapes, as bottles, images, etc., and the rubber is sometimes exported in these uncouth forms.

† Mr. Goodyear had been experimenting to find some way of rendering rubber insensible to heat and cold. It is said that one day, while talking with a friend, he happened to drop a bit of S in a pot of melted rubber. By one of those happy intuitions which seem to come only to men of genius, he watched the process, and to his amazement found that while the appearance of the rubber was the same—elastic, odorous, and tasteless—its stickiness was gone, and it had gained the properties he so much desired. He immediately took out a patent in this country and sailed for England, where, instead of securing his secret by a similar patent, he offered to sell it for £10,000. Charles Hancock, with whom he had been corresponding for several years, and who had been engaged in similar experimenting, resolved to discover it himself. He shut himself up in his labo-

by cold or softened by heat, and admits of many uses to which common rubber would be entirely unsuited. If sulphurized rubber be heated to a high temperature it becomes a hard, brittle, black solid, capable of a high polish, which is used for knife-handles, combs, buttons, etc.

Gutta-percha ($C_{20}H_{32}$) resembles caoutchouc in its source, preparation, and appearance. It softens in warm water, and can then be moulded like wax. When cooled it assumes its original solidity. It is extensively used in taking impressions of medals, etc.

THE ALBUMINOUS BODIES.

THESE are albumen, casein, gelatin, and fibrin. Owing to the complexity of their composition, no satisfactory formula can be assigned to them. The molecule of albumen has been stated as $C_{72}H_{110}N_{18}SO_{22}$, but it is very uncertain.*

Albumen is found nearly pure in the whites of eggs †—hence the name (*albus*, white). It exists in two amorphous conditions—as a liquid in the sap of plants, the humors of the eye, serum of the blood, etc.; and as a

ratory and went to work. Disheartening failures marked every attempt. At last he tried S. At first, he did not succeed; but, persevering, he finally saw, amid the stifling fumes of brimstone, the soft rubber metamorphosed into the vulcanized caoutchouc. He, too, was possessed of the secret, and, taking out a patent, reaped the reward of his patient labor.

* Many chemists regard albumen, casein, fibrin, etc., as chemically identical and capable of being converted by the vital force one into the other. These bodies are sometimes called *Protein* (*protos*, first) on the supposition that they were derived from a single azotized principle named protein.

† Strange to say, "the venom of the rattlesnake is isomeric with the 'whites of eggs.'"

solid in the seeds of plants, and the nerves and brains of animals.*

Properties.—It is soluble in cold, but insoluble in hot H_2O. At a temperature of about 140° F. it coagulates. This change we always see in the cooking of eggs; yet nothing is known of its cause. Alcohol, corrosive sublimate, acids, creosote, and solutions of copper, lead, silver, etc., have the power to coagulate albumen. In cases of poisoning by these substances, the white of eggs is therefore a valuable antidote, as it wraps them in an insoluble covering, and so protects the stomach.

Casein (*caseus,* cheese) is found in the curd of milk. In the presence of an acid it coagulates, and thus milk curdles after it sours. Rennet (the dried stomach of a calf) is used to coagulate milk in the process of cheese-making, but the cause of its action is not understood.

Milk is a natural emulsion, composed of exceedingly minute globules diffused through a transparent liquid. The globules consist of a thin envelope of casein filled with butter. Being a trifle lighter than H_2O, they rise to the surface as cream. Churning breaks these coverings, and gathers the butter into a mass. Milk contains some sugar, which by a peculiar change termed "lactic fermentation" is converted into lactic acid. The casein seems to act as a ferment in hastening this oxidation, and by its decay produces the offensive odor. In the "sour-

Fig. 73.

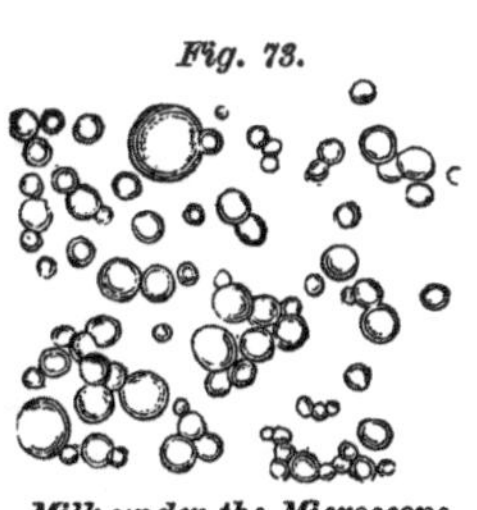

Milk under the Microscope.

* This principle is of very great importance, as albumen may thus be carried by the blood through the system, but when once deposited it cannot be dissolved and washed away again.

ing" of milk there is no extrication of gas and no absorption of O. The milk-sugar ($C_{12}H_{24}O_{12}$) disappears and lactic acid ($C_3H_6O_3$) gradually takes its place. It is an excellent illustration of a complex molecule breaking up into simple ones.

Gelatin.—Hot water dissolves a substance from animal membranes, skin, tendons, and bones,* which, on cooling, forms a yielding, tremulous mass called gelatin. In calves-foot jelly, soups, etc., it is well known.† *Glue* is a gelatin made from bones, hoofs, horns, etc., by boiling in H_2O and then evaporating the solution. *Isinglass* is a very pure gelatin, obtained from the air-bladders of the cod, sturgeon, and other fish. *Size* is a gelatin prepared from the parings of parchment. It is used for sizing paper in order to fill up the pores and prevent the ink from spreading, as it does on unsized or blotting-paper.

Fibrin constitutes chiefly the fibrous portion of the

* Bones consist of organic and mineral matter combined.

ANALYSIS. (*Berzelius.*)

Gelatin	32.17
Blood-vessels	1.13
Phosphate of lime	51.04
Carbonate of lime	11.30
Fluoride of calcium	2.00
Phosphate of magnesia	1.16
Chloride of sodium	1.20
	100.00

By soaking a bone in HCl the mineral matter will all be dissolved, and the organic matter left in the original shape of the bone, but soft and pliable. If, instead, the bone be burned in the fire, the organic matter will be removed and the mineral left white and porous. (See *Physiology*, p. 20.)

† As an article of food it is of very little nutritive value. It may answer to dilute a stronger diet, but of itself does little to build up the body of an invalid. Beef-tea, even, is now thought to have little nourishing property, its principal office being to act as a stimulant.

Fig. 74.

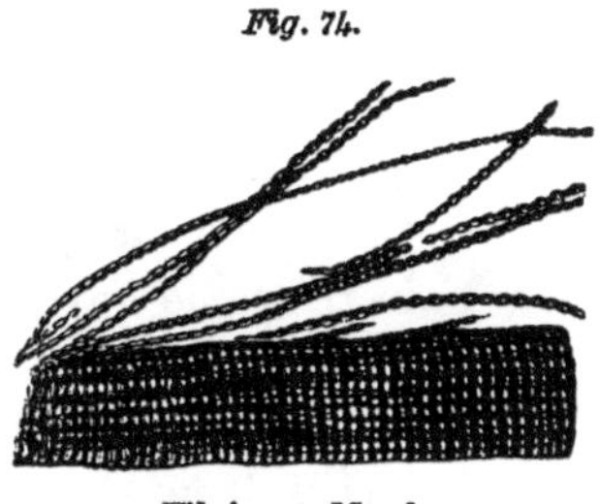

Fibrin, or Muscle.

muscles. If a piece of lean beef be washed in clean H_2O until all the red color disappears, the mass of white tissue which will remain is called *fibrin*. Like albumen, it exists in two forms—as a liquid in the blood, and as a solid in the flesh. The clotting of blood is due to the coagulation of the fibrin. (See *Physiology*, p. 108.)

Vegetable Albuminoids.—Vegetables contain substances which are scarcely to be distinguished from the albuminous bodies derived from animal sources. If wheat flour be made into a dough, and then kneaded in water until the soluble portion is washed away, the tough, sticky mass which will remain is called *gluten*. It is a nitrogenous substance, allied to albumen. It exists most abundantly in the bran of cereal grains.

By treating peas as we do potatoes in forming starch, and then adding a little acid to the water which is left after the starch settles, an albuminous substance is deposited, which is thought to be identical with casein. The Chinese use it largely for cheese. It is found abundantly in the seeds of peas, beans, etc., and is termed *legumin*.

Putrefaction.— Owing to the complex structure of albuminous substances, and the presence of N, they readily oxidize and form new and simple compounds. This breaking up of the organic structure is called putrefaction. Any albuminous substance thus putrefying may act as a ferment. This probably explains the danger physicians incur in dissecting a dead body. The least portion of the decomposing matter entering the flesh, through a scratch even, is liable to be fatal. The absence of H_2O retards chemical change, and therefore, meats, apples, etc., are preserved by drying.* Salt acts somewhat in the same way by absorbing the juice of the meat, and, while it covers it as brine, wards off the attacking O; but as it dissolves some of the salts and other valuable elements, it makes the meat less nutritious.

DOMESTIC CHEMISTRY.

In the chemistry of housekeeping there are some points not yet mentioned, which may now be profitably discussed.

Making Bread.—Flour consists of gluten, starch, and a little dextrine and sugar.

The oily matter and the salts—of which there are from one to two per cent. in wheat—are contained mainly in the bran. The process of mixing the "sponge" is purely mechanical. When the sponge is set in a warm place to rise (as heat favors chemical change), the yeast, yeast-cake, or emptyings,† as the case may be, induces a rapid

* The cold also protects from chemical change. The bodies of mammoths have been found in the frozen soil of the Arctic regions so perfectly preserved that the dogs ate the flesh. How long the animals had been there we cannot tell, but certainly for ages. In 1861 the mangled remains of three guides were found at the foot of the Glacier de Boisson, in Switzerland. They had been lost in an avalanche on the plateau of Mont Blanc, forty-one years before.

† Milk-emptyings are sometimes used in making bread. In this case, the mixture of flour and milk, kept at a temperature of about "blood heat," rapidly develops yeast, which produces fermentation. If the heat is much above this, the plant will be killed, and the milk be merely turned to lactic acid. Oftentimes, too, the side of the dish, near the fire, may be warm enough to produce yeast and to generate CO_2 and alcohol, while on the opposite side lactic acid is being formed. A uniform temperature is necessary, and this can best be ob-

fermentation, converting the sugar into alcohol and CO_2. This gas is diffused through the mass, and being retained by the tenacious and viscid dough, causes it to "rise," *i. e.*, to swell and become porous. The next step includes the addition of fresh flour, and a laborious process of "kneading." The latter, so essential to good bread, diffuses the half-fermented sponge uniformly through the dough; it also breaks up into smaller ones the bubbles of gas entangled in the gluten, and thereby makes the bread fine-grained. The dough is now "moulded" into loaves, and then placed in the oven. The heat rapidly expands the CO_2, and increases the porosity of the bread. The starch granules are broken up and the alcohol vaporized, and, with a part of the H_2O, driven off. The surface gradually becomes dry and hard, and losing a part of its chemically combined water, is partially converted into a substance allied to caramel, thus forming the crust.* If the temperature of the oven is right, the cells of the bread will have sufficient strength to retain their form after the gas and vapors have escaped. If the heat is not sufficient, or if there is too much water in the dough, the CO_2 escapes, the cells, not being sufficiently hardened, collapse, and the bread is "slack-baked." If the oven is too hot, the crust forms too quickly over the surface of the loaf, preventing the escape of the CO_2, which accumulates at the centre, making the bread hollow.

Stale Bread.—New bread consists of nearly one-half water. In stale bread this disappears. It has, how-

tained by placing the dish of emptyings in a kettle of warm water on the stove hearth.

* A shiny coat is given to the loaf ("rusk") by moistening the crust after the bread is baked, thus dissolving some of the dextrine, which is also contained in the crust. This quickly dries on returning it to the oven.

ever. only chemically combined with the solid portions, and may be brought to view by heating the loaf in a close tin vessel.

Aerated Bread is made in the following manner: Flour and salt are put in a revolving copper globe, into which H_2O charged with CO_2 is admitted. When well mixed, a stop-cock is turned and the dough is driven out, by the elastic force of the gas, into pans ready for baking.

Sour Bread results from a neglect to arrest the first stage of the fermentation, thus allowing the second stage to commence and acetic acid to be formed. The acid is neutralized by an alkali, as saleratus, or soda.

Griddle-cakes are raised by the addition of some ferment, as yeast; but the second, or acetic stage, is always reached. The "batter" then tastes sour, and is sweetened by saleratus or soda. The acetic acid combines with the metallic base, forming a harmless salt which remains, while the CO_2 bubbles up through the batter, making it "light."

Raising Biscuit.—In raising biscuit or cake, soda and cream of tartar* are most commonly used. The CO_2 is set free, and, escaping as a gas, makes the dough porous, while the sodium and potassium tartrate (Rochelle salt) which is left is a simple salt. Ordinary "baking-powders" are merely cream of tartar and soda. A variety invented by Prof. Horsford contains acid calcium phosphate (see note, p. 140); this reacting upon the "soda" forms calcium and sodium phosphates, both of which are

* Cream of tartar is often adulterated with plaster, lime, chalk, or flour. By dissolving in water, these impurities can be detected, as they form an insoluble precipitate; but in milk as commonly used in cooking, they are not noticed.

materials for bone-making.* Soda and HCl are also used in baking. By heat both constituents are resolved into H_2O, CO_2, and NaCl. The H_2O and CO_2 raise the bread, while the common salt seasons it. There is a difficulty in procuring pure acid and in mixing the ingredients in their combining proportions. Sal-volatile (ammonium sesquicarbonate, p. 135) is often used by bakers for raising cake. This should volatilize into two gases, H_3N and CO_2, on the application of heat, but in practice a portion is commonly left hidden in the cake, and may be detected by the odor. Alum is often employed by bakers to whiten bread and render the gluten of inferior flour more tenacious.

Toasting Bread. — By toasting, bread becomes much more digestible, as the starch is converted largely into dextrine, which is soluble. The charcoal which may be formed when the heat has disorganized the bread and driven off the water, also acts favorably on the stomach by absorbing in its pores noxious gases, as in "crust-coffee."

Cooking Potatoes.—A raw potato is indigestible, but by cooking, the starch granules absorb the water of the potato, burst, and make it "mealy." If the potato contains more H_2O than the starch can imbibe, it is called "watery."

* It is doubtful whether ordinary yeast-powders or cream of tartar and soda make as healthy food as the regular process of fermentation. There is frequently a portion of the powders left uncombined, and always a salt formed which may perhaps interfere with the action of the gastric juice. Sometimes, indeed, we find biscuit and cake yellow, and even spotted with bits of saleratus; yet, through a false economy, such food is too often "eaten to save it."

PRACTICAL QUESTIONS.

1. How would you prove the presence of tannin in tea?
2. How would you test for Fe in a solution?
3. Why can we settle coffee with an egg?
4. How would you show the presence of starch in a potato?
5. Why is starch stored in the seed of a plant?
6. Why are unbleached cotton goods dark-colored?
7. Why do beans, rice, etc., swell when cooked?
8. Why does decaying wood darken?
9. Why does smoke cure hams?
10. How would you show that C exists in sugar?
11. Why do fruits lose their sweetness when over-ripe?
12. Why does maple-sap lose its sweetness when the leaf starts?
13. Should yeast cakes be allowed to freeze?
14. Why will wine sour if the bottle be not well corked?
15. Why can vinegar be made from sweetened water and brown paper?
16. Why should the vinegar-barrel be kept in a warm place?
17. Why does "scalding" check the "working" of preserves?
18. Is the oxalic acid in the pie-plant poisonous?
19. How may ink-stains be removed?
20. Why is leather black on only one side?
21. Why do drops of tea stain a knife-blade?
22. Why will not coffee stain it in the same way?
23. Why does writing-fluid darken on exposure to the air?
24. What causes the disagreeable smell of a smoldering wick?
25. Why does ink corrode steel pens?
26. How does a bird obtain the $CaCO_3$ for its egg shells?
27. Why will tallow make a harder soap than lard?
28. Why does new soap act on the hands more than old?
29. What is the shiny coat on certain leaves and fruits?
30. Why does turpentine burn with so much smoke?
31. Why is the nozzle of a turpentine bottle so sticky?
32. Why does kerosene give more light than alcohol?
33. What is the antidote to oxalic acid? Why?
34. Would you weaken camphor spirits with water?
35. What is the difference between rosin and resin?
36. Why does skim-milk look blue and new milk white?
37. Why does an ink-spot turn yellow after washing with soap?

CONCLUSION.

Chemistry of the Sunbeam.—The various plant-products of which we have spoken in Organic Chemistry, when burned, either in the body as food or in the air as fuel, give off heat. This was garnered in the plant while growing, and came from that great source of heat—the sun. Thus all vegetation contains the latent heat of the sunbeam, ready to be set free upon its own oxidation. The coal, even, derived as it is from ancient vegetation, hidden away in the earth, is thus a mine of reserved force. Those black diamonds we use as fuel become, in the eye of science, crystallized sunbeams, fagots of force, ready to impart to us at any moment the heat of some old Carboniferous day. A field of growing wheat reaches out its tiny arms, and tangling in stalk and grain the heat of sultry mid-summer, retains it against the bleak December. The oil-well spouts not alone unsavory kerosene, but liquid sunbeams, the gathered store of a geologic age. As we warm ourselves by our fires, or sit and read by our oil and gas lights, how strange the thought that their light and heat streamed down upon the earth ages ago, were absorbed by the grotesque leaves of the old coal forests, and kept safely stored away by a Divine care, in order to provide for our comfort! The present warmth of our bodies all came from the same source—the sun. It mostly fell in the sunbeams of last summer upon our gardens and fields, was preserved in the potatoes, cabbage, corn, etc., we have eaten as food, and to-day reappears as heat and motion. Every blow, every breath, and every step, are but transformations of solar rays and can be estimated in sunshine.

The Sun the Source of Power.—The Sun warms, enlivens, and animates the earth. In the laboratory of the leaf he produces the most wonderful chemical changes. We see his handiwork in the building of the forest, the carpeting of the meadow, and the tinting of the rose. On the ladder of the sunbeam water climbs to the sky, and falls again as rain. The very thunder of Niagara is but the sudden unbending of the spring that was first coiled by the sun in the evaporation from the ocean. Up to the sun, then, we trace all the hidden manifestations of power. Yet the force that produces such intricate and wide-extended changes is only one twenty-three hundred millionth part of the tide that flows in every direction from this great central orb. But what is our sun itself save a twinkling star beside great suns like Sirius, and Regulus, and Procyon, whose brilliancy in the far-off regions of space drowns our little sun as the dazzling light of day does the smouldering blaze of some wandering hunter?

Changes of Matter.—Chemical changes are taking place wherever we look—on land or sea. The hard granite crumbles and moulders into dust. The stout oak draws in the air and solidifies it; takes up the earth and vitalizes it; changes all into its own structure, and proudly stands monarch of the forest. But in time its leaves turn yellow and sere; its branches crumble; itself totters, falls, and disappears. Our bodies seem to us comparatively stable, but, with the rock and the oak, they too pass away. All Nature is a torrent of ceaseless change. We are but parts of a grand system, and the elements we use are not our own. The water we drink and the food we eat to-day may have been used a thousand times before, and that by the vilest beggar or the lowest earth-worm. In Nature all is common, and no use is base. Those particles of matter we so fondly call our own, and decorate so carefully, a few months since may have dragged boats on the canal, or waved in the meadow as grass or corn.* From us they will pass on their ceaseless round to develop other forms of vegetation and life, whereby the same atom may freeze on arctic snows, bleach on torrid plains, be beauty in the poet's brain, strength in the blacksmith's arm, or beef on the butcher's block. Hamlet must have been somewhat more of a chemist than a madman when he gravely assured the king that "man

* The truth that matter passes from the animal back to the vegetable, and from the vegetable to the animal kingdom again, received, not long since, a curious illustration. For the purpose of erecting a suitable monument in memory of Roger Williams, the founder of Rhode Island, his private burying-ground was searched for the graves of himself and wife. It was found that everything had passed into oblivion. The shape of the coffins could only be traced by a black line of carbonaceous matter. The rusted hinges and nails, and a round wooden knot, alone remained in one grave; while a single lock of braided hair was found in the other. Near the graves stood an apple-tree. This had sent down two main roots into the very presence of the coffined dead. The larger root, pushing its way to the precise spot occupied by the skull of Roger Williams, had made a turn as if passing around it, and followed the direction of the backbone to the hips. Here it divided into two branches, sending one along each leg to the heel, when both turned upward to the toes. One of these roots formed a slight crook at the knee, which made the whole bear a striking resemblance to the human form. (These roots are now deposited in the museum of Brown University.) There were the graves, but their occupants had disappeared; the bones even had vanished. There stood the thief—the guilty apple-tree—caught in the very act of robbery. The spoliation was complete. The organic matter—the flesh, the bones, of Roger Williams—had passed into an apple-tree. The elements had been absorbed by the roots, transmuted into woody fibre, which could now be burned as fuel, or carved into ornaments; had bloomed into fragrant blossoms, which had delighted the eye of passers-by, and scattered the sweetest perfume of spring; more than that—had been converted into luscious fruit, which, from year to year, had been gathered and eaten. How pertinent, then, is the question, "Who ate Roger Williams?"

may fish with the worm that hath eat of a king, and eat of the fish that hath fed of the worm."

Shakespeare expresses the same chemical thought when he says:

"Imperious Cæsar, dead and turned to clay,
Might stop a hole to keep the wind away.
Oh! that the earth which kept the world in awe
Should patch a wall to expel the winter's flaw!"

Or, again, when he makes Ariel sing:

"Full fathom five thy father lies:
Of his bones are coral made;
Those are pearls that were his eyes;
Nothing of him that doth fade
But doth suffer a sea change
Into something rich and strange."

Life and Death are thus throughout nature commensurate with and companions of each other. Oxygen is the destroyer, and the sunbeam the builder. Oxygen tears down every living structure, and would bring all things to rest in ashes. The sunbeam re-invigorates, rebuilds, and rescues from the grasp of decay. Though they seem to be antagonists, oxygen and the sunbeam really work in harmony, and each supplements the labor of the other. Death alone makes life possible.

Thus we have traced some of the wonderful processes by which this world has been arranged to supply the varied wants of man. Wherever we have turned, we have found proofs of a Divine care planning, conforming, and directing to one universal end, while from the commonest things and by the simplest means the grandest results have been attained. Thus does Nature attest the sublime truth of Revelation, that in all, and through all, and over all, the Lord God omnipotent reigneth.

IV.

Appendix

NAMES OF CHEMICALS

ACCORDING TO

THE OLD AND THE NEW NOMENCLATURE

	THE NEW.	THE OLD.
1.	Ammonium carbonate	Carbonate of ammonia.
2.	" chloride	Chloride of ammonium.
3.	" sulphate	Sulphate of ammonia.
4.	Antimony sulphide	Sulphide (sulphuret) of antimony.
5.	Barium sulphate	Sulphate of baryta, or Barytes
6.	Calcium carbonate	Carbonate of lime.
7.	" chloride	Chloride of calcium.
8.	" hypochlorite	Hypochlorite of lime.
9.	" oxide	Lime.
10.	" phosphate	Phosphate of lime.
11.	" sulphate	Sulphate "
12.	" sulphite	Sulphite "
13.	Carbon disulphide	Bisulphide (bisulphuret) of carbon.
14.	Carbonic anhydride*	Carbonic acid.
15.	Copper nitrate	Nitrate of copper.
16.	" oxide	Oxide "
17.	" sulphate	Sulphate "
18.	Ferric oxide	Sesquioxide of iron.
19.	" hydrate	Hydrated sesquioxide of iron.
20.	Ferric disulphide	Bisulphide (bisulphuret) of iron.
21.	Ferrous sulphide	Sulphide (sulphuret) of iron.
22.	Hydrogen potassium carbonate	Bicarbonate of potash (potassa).
23.	" protoxide (water, H_2O)	Protoxide of hydrogen (HO).
24.	" sodium carbonate	Bicarbonate of soda.
25.	" sulphide	Sulphide of hydrogen.
26.	Hyponitric anhydride (acid)	Nitrous acid.
27.	Iron disulphide	Bisulphide (bisulphuret) of iron.
28.	" sulphide	Sulphide (sulphuret) of iron.
29.	" sulphate	Sulphate of iron, or Protosulphate of iron
30.	Lead acetate	Acetate of lead.
31.	" carbonate	Carbonate of lead.
32.	" oxide	Oxide of lead.

* See note, p. 29. In the old nomenclature it is customary to apply the term acid indifferently to the hydride (hydrous) or anhydride (anhydrous).

	The New.	The Old.
33.	Lead silicate	Silicate of lead.
34.	" sulphide	Sulphide (sulphuret) of lead.
35.	Magnesium oxide	Magnesia.
36.	" carbonate	Carbonate of magnesia, or Magnesia.
37.	" sulphate	Sulphate of magnesia.
38.	Manganese dioxide	Binoxide of manganese.
39.	Mercuric chloride	Chloride of mercury.
40.	Mercurous chloride	Subchloride of mercury.
41.	Mercuric oxide	Red oxide of mercury.
42.	Mercury sulphide	Sulphide (sulphuret) of mercury.
43.	Nitric anhydride	Anhydrous nitric acid.
44.	Potassium bromide	Bromide of potassium.
45.	" carbonate	Carbonate of potash.
46.	" chlorate	Chlorate of potash.
47.	" chloride	Chloride of potassium.
48.	" cyanide	Cyanide of potassium.
49.	" ferricyanide	Ferricyanide of potash.
50.	" ferrocyanide	Ferrocyanide of potash.
51.	" hydrate	Hydrated potash (potassa), or Potash.
52.	" iodide	Iodide of potassium.
53.	" nitrate	Nitrate of potash.
54.	" permanganate	Permanganate of potash.
55.	" sulphate	Sulphate of potash.
56.	Silicic anhydride	Anhydrous silicic acid.
57.	Silver chloride	Chloride of silver.
58.	" cyanide	Cyanide of silver.
59.	" iodide	Iodide of silver.
60.	" nitrate	Nitrate of silver.
61.	" sulphate	Sulphate of silver.
62.	Sodium biborate	Biborate of soda.
63.	" carbonate	Carbonate of soda.
64.	" chloride	Chloride of sodium.
65.	" hyposulphite	Hyposulphite of soda.
66.	" nitrate	Nitrate of soda.
67.	" phosphate	Phosphate of soda.
68.	" silicate	Silicate of soda.
69.	" sulphate	Sulphate of soda.
70.	Sulphuric anhydride	Anhydrous sulphuric acid, or Sulphuric acid.
71.	" acid	Hydrated sulphuric acid, or Sulphuric acid.
72.	Sulphurous anhydride	Anhydrous sulphurous acid.
73.	Tin oxide	Oxide of tin.
74.	Zinc oxide	Oxide of zinc.
75.	" Sulphate	Sulphate of zinc.

Directions for Experiments.

The following simple suggestions will enable any student to perform all the experiments mentioned in this work. Many easy illustrations are also given in addition to those named in the text. Carefully compare them with those contained in the body of the book. The Italic figures refer to the pages of the book, and the small ones to the number of the experiment.

18.—1. Put into the mortar as much potassium chlorate as will lie upon the point of a knife-blade, and half as much sulphur. Cover the mortar with a sheet of writing paper, having a hole cut in it just large enough for the handle of the pestle to pass through. When the two substances have become thoroughly mixed, grind heavily with the pestle, when rapid detonations will ensue. The paper will prevent loose particles from flying into the eyes. The same precaution should always be observed when pulverizing potassium chlorate. A better way is to purchase that salt in powder, or to make a hot saturated solution and pour it out in thin films on panes of glass or old plates. It then forms very small crystals, which can be scraped off and dried for use. After using, clean out the mortar carefully for other experiments. The powder can be wrapped with paper into a hard pellet and exploded on an anvil by a sharp blow from a hammer. Sometimes small bits of phosphorus are used instead of sulphur. Great care is then necessary, as the particles of burning phosphorus are apt to fly to some distance.

2. Dissolve 40 grs. of common soda in one wine-glass of water, and 35 grs. of tartaric acid in another. On being poured together in a goblet they will violently effervesce. Use a glass which is large enough to prevent any of the liquid from running over. Neatness in experiments is essential to perfection and often to success. At the close of this illustration, evaporate the solution,* and a neutral salt will result (page 211).

22-3.—1. A few drops of vinegar, or any acid, will turn the purple cabbage-solution to a bright red ; and a little of the potash solution, to a deep green. Add alcohol to the red solution to keep it from freezing, and bottle for use. Dissolve

* Pour a part of the liquid into an evaporating dish, and place this on the tripod over the flame of the spirit-lamp,or upon a hot stove. Heat until a drop of the liquid taken out on the end of a glass rod and put on a bit of glass will crystallize as soon as it cools. Then set the dish aside to cool, when crystals will soon begin to form.—In this connection it is well to remark that a *cook-stove will be found of great use in chemical experiments, and indeed may, in the laboratory, take the place of a furnace.* The oven will dry apparatus and chemicals ; the heat is sufficient for evaporating solutions, distilling water, etc., while an excellent sand or water bath may be readily contrived.

20 or 30 grs. of litmus in an oz. of water; filter and bottle. Dissolve also a stick of potash in 4 oz. of water; filter and bottle. Fill two test-tubes nearly full of water; color one with the cabbage solution and the other with the litmus solution. To each add alternately a few drops of the potash solution and of oil of vitriol. The color can be changed at pleasure.

A pipette—a glass-tube with a bulb in the middle and one end drawn to a point—will be found convenient for dropping liquids. In lieu of this, take a piece of glass-tubing,* and heating the end in the flame of the spirit-lamp (the greatest heat is near the tip), seal the openings. This tube will readily take up a drop upon its extremity, and will be found useful for stirring liquids.

27.—1. Put into a dry test-tube 24½ grains of pure potassium chlorate and heat cautiously. The test-tube may be supported by a strip of thick paper twisted around it at the top. Move the tube to and fro through the flame at first, until it becomes fully heated; keep the tube inclined and not perpendicular, letting the flame strike the side rather than the bottom. Hold the thumb lightly over the mouth of the tube. The salt melts quietly and the tube will soon be filled with O, as proved by its re-igniting a glowing splint introduced into it. When no more gas is evolved, allow the salt to cool, shaking it gently to prevent its attaching itself to the tube. When cold the residue will weigh 14.9 grains. The residue is no longer a *chlorate*, and it gives out no yellow gas if moistened with H_2SO_4, but will yield a white precipitate with $AgNO_3$, which shows it to be a *chloride*. Most of the following experiments may be performed in test-tubes as above, but when it is desirable to make a larger quantity of O, one ounce of potassium chlorate is *very carefully* pulverized and mixed with half that quantity of black oxide of manganese, which has just been strongly heated on an iron plate or sand-bath, and allowed to cool.† Be careful not to grind them together. The mixing is effected by placing them both on sheets of clean paper and pouring them back and forth from one sheet to the other until the mixture has a uniform gray color. Place the mixture in a flask; fit a cork to the nozzle; then withdraw the cork, and with a round file bore a hole through it just large enough ‡ to admit a glass tube bent § as shown in Figs. 2 or 10. Return the cork

* The tube may be cut of any length. Lay it upon the table, and with a three-cornered file make a deep scratch where you wish to break it; then hold the tube in both hands, placing a thumb on each side of the scratch, and with a steady pressure the glass will break at the desired point. Two tubes, each closed at one end, may also be easily obtained by heating a piece of tubing in the flame until a ring of the glass becomes very soft, when, by pulling upon the opposite ends of the tube, the heated portion will be drawn out, diminished in size, and the opening closed. A little practice will enable the student to do this neatly and expertly. Gas-jets may also be made in this way for the experiments illustrated in Figs. 12, 15, 16, and 19.

† In order to test the purity of the materials, and thus avoid any danger of an explosion, it is well, previous to putting the mixture in the flask, to place a little in an iron spoon and heat it over the lamp. If the gas pass off quietly, no danger need be apprehended.

‡ Whenever corks leak gas they may be wrapped with thin strips of wet paper to make them fit more tightly; or the entire nozzle may be smeared with tallow, or covered with sealing-wax, if heat is not used. In that case use a poultice of linseed meal, or a little plaster-of-Paris may be wet and applied.

§ A glass tube may be bent at any point by softening that part in the flame of the spirit-lamp. Practice alone will give the required expertness. The following points should be observed: 1. Keep the tube constantly turning between the fingers so that it may be equally heated on all sides; 2. Do not twist or pull

and tube, arrange the apparatus as shown in either of the figures, and apply the heat. This must be done very cautiously at first, holding the lamp in the hand and moving it around so that the flame may strike all the lower part of the flask, and thus expand it uniformly. Be careful also that no draft of cold air strikes against the heated glass. The first few bubbles of gas will consist mainly of the air contained in the flask, and should not be caught. When the gas begins to pass over freely, diminish the heat.* *When the gas ceases, remove the stopper from the flask, or lift the end of the tube out of the water;* otherwise, as the flask cools, and a vacuum is formed, *the water in the tub will rush back into the flask and break it.* When the retort is nearly cool, pour in some warm water to dissolve the residuum, which may then be poured out and the flask dried for future use.

Instead of bending the glass tubing it may be cut into short lengths and the pieces joined by bits of rubber tubing, as in Figs. 10 and 13. The advantage of this is that the flexible joints are not liable to break, and the apparatus may be more easily moved. Where a large quantity of O is to be made, a copper retort and rubber tubing will be found cheap and convenient. No especial care is then needed in managing the heat.†—In place of the pneumatic tub, a pail or a tin pan may be used, letting the bottle rest on a shelf as in Fig. 10, or on a couple of bricks.—The bottles for collecting the gas may be the regular "deflagrating jar" of the chemist, or the common "packing bottle" of the druggist. They are to be sunk in the water of the pneumatic tub and filled; then inverted and lifted upon the shelf, carefully keeping the lower edge of the bottle under the water. The bottles may also be filled from a pitcher, then closed with the hand or a plate, and quickly inverted and placed on the shelf in the tub or pan ready for use. As soon as a bottle is filled with gas a plate may be slipped under the mouth, and thus, leaving enough water in the plate to cover the lower edge, be set aside as in Fig. 2. Gas may be passed from one jar to another in the manner shown in Fig. 17.—While the gas is being collected, the water from the bottles which are filling may cause the tub to overrun; to prevent this, arrange a siphon to carry off the water into a pail below the table.—When a jar of gas is wanted for use, remove it to the tub, slip a plate under the mouth, or simply close it with the hand, and lifting the jar out, carry it to the table and place it mouth upward. Uncover only when the experiment is ready to be performed, as the gas will slowly diffuse.

29–31.—1. The experiment with the candle may be very strikingly performed by filling a common fruit-jar with O, and another with N. The covers may be loosely laid on top, and the lighted candle passed quickly from one to the other, as mentioned in note on page 43. The candle may be simply stuck on the end of a bent wire, as in Fig. 14, but it is much neater to have the tinsmith fit a little cup for its reception, as shown in the figure.

the tube while heating; 3. Do not bend it until very soft; if not hot enough the elbow will be flattened. Blowing gently into the tube at the moment of bending also prevents flattening.

* The gas often looks cloudy, owing to little particles of the salt which are carried over suspended in it in fine powder; but these gradually become dissolved in the water.

† If, during the operation, the gas suddenly ceases to come off, remove the flame and ascertain whether the delivery tube is not choked up, which would result in a very violent explosion of the retort.

2. Worn-out watch-springs can be obtained gratis of any jeweller, and may be easily straightened by slightly heating and then drawing them between the fingers. If the end of each spring be strongly heated and then pounded with a hammer on any smooth, hard surface, the temper may be drawn and the edge sharpened. Make a slit with a knife in the side of a match, into which insert the edge of the spring. Take a piece of zinc or tin large enough to cover the mouth of the jar containing the O, and make a hole through it with a nail. Pass the other end of the spring through this hole, and then through a thin cork. The spring is now ready for burning. The metal cover will prevent the flame from coming out of the jar and burning one's hand, and the cork will hold the spring in its place.* When the match is ignited, and then lowered into the jar of O, the spring should not reach more than half-way to the bottom, and should be pushed down as it burns. A cheap packing-bottle should be used, as the glass is frequently broken by the melted globules of iron. Do not fill it quite full of gas, as then, on inverting, a little water will be left at the bottom, or some fine sand may be thrown into the jar before the experiment. The illustration may be repeated with a coil of fine iron wire. The springs from an old hoop-skirt burn nicely in O.

3. If brimstone be used in the experiment with S, and it fails to light readily, pour upon it a few drops of alcohol, and then ignite.

4. When S, P, charcoal, a wax candle, Na, and other substances are to be burned in O, they may be supported on the end of pieces of glass tube bent like the letter J, and left open at the shorter end only. For this purpose, covers must be provided with holes large enough for the glass tube to pass through. Or a "deflagrating spoon" may be readily extemporized to contain the phosphorus. Hollow a small piece of chalk and attach a wire to it, which may then be secured to a metal top, as in the case of the watch-spring. This need not be pushed down into the jar as the burning progresses. Be careful to cut the phosphorus under water, to dry it carefully with blotting-paper, and not to handle it. The fumes are very disagreeable, and should not be inhaled or allowed to escape into the room. They soon dissolve when shaken with a little water, which then gives an acid reaction, H_3PO_4.

5. The experiment of burning P under water may be easily shown by throwing a piece of P into a glass of hot water. It melts and looks like a thick oil under the water. A fine stream of O gas may then be passed through a glass tube down to the P, which burns brightly under water. (Note, p. 130.)

6. The oxidation of one solid by means of nascent O liberated from another solid can be shown as follows: Heat about 5 grams of potassium nitrate (saltpetre) in a test-tube until it melts quietly. Remove the lamp and throw in pieces of S as large as peas, when they burn with an intensely bright flame. The heat is often sufficient to melt the glass, and the precaution should be taken to hold it over an iron plate or sand-bath. Or melt a quarter of a pound of saltpetre in an evaporating dish; an ordinary tin cup will answer. Put it on some burning coals in a draught to carry off the fumes. Plunge into the liquid a piece of bark-

* It is well to obtain several pieces of thin board or shingles, about 6 inches square, bore a small hole in the center and insert a match end or small plug. These may be used as covers in most experiments where a deflagrating spoon is to be employed. The handle of the spoon is passed through this hole and held in place by the small plug.

charcoal strongly ignited. The oxygen of the saltpetre will support the combustion, and the charcoal will deflagrate in a rushing volcano of scintillations.

7. In burning bark-charcoal in O, force the gas into the bottle through a bit of rubber-tubing at the mouth. By placing this at one side, the gas is given a rotary motion, and the sparks of ignited charcoal will drive around the bottle in a beautiful maelstrom of fire. The gas may be forced in from a rubber bag, and this striking effect easily produced.

8. Arrange a receiver upon the bed-plate of the air-pump so that O may be admitted from a gas-bag by turning a stop-cock. Put under the receiver an ignited tallow candle with a big wick. Exhaust the air until the flame goes out and there is left only a coal of fire. Admit the air, and it will have no effect to restore the blaze. Force in some O quickly, and the coal will burst instantly into a brilliant white light, brighter than at the first.

9. Burn a small piece of Na or K in a jar of O, and dissolve the white powder formed. The solution restores reddened litmus, showing it to be an alkali.

39.—1. Put in an evaporating dish a little starch; cover it with water in which a few crystals of potassium iodide have been dissolved, and heat. Stir the liquid, to prevent lumps. When cooked, immerse in the paste slips of white blotting or clean writing-paper, and hang them up to dry. They must be moistened when used. Be careful not to heat the glass tube too hot, lest the ether-vapor may ignite. Keep the jar well filled with vapor by frequently shaking it. Lower into the ozone a bit of silver-leaf moistened with water; it will quickly crumble into the oxide.

2. Ozone may also be prepared by the slow oxidation of phosphorus in the following manner: Scrape off the white coating of a stick of phosphorus under water, and cut the cleansed phosphorus into pieces 12 or 15 millim. long.* Place one of these pieces in a wide-mouth litre-bottle full of air, with about a teaspoonful of water at the bottom. Close the mouth of the bottle with a glass plate, and expose the whole for an hour or two to a temperature of 15° or 20° C. Then invert the neck of the bottle in water, and allow the phosphorus to fall out. Replace the glass plate, and withdraw the bottle and its contents from the water. The phosphorus in this experiment undergoes a slow oxidation, during which a little ozone is formed, and is left mixed with the air; but the ozone will be again destroyed if it is left too long with the phosphorus.

3. Add to the bottle of air which has been ozonized by means of phosphorus, a few drops of a *very* dilute blue solution, formed by dissolving powdered indigo in strong sulphuric acid, and then diluting it with water. If the blue liquid be shaken up with the ozonized air, the color will quickly disappear.

4. In making ozone in the ordinary way, with a hot glass rod stirred in ether-vapor, when there is water at the bottom of the jar, some peroxide of hydrogen will be made and absorbed by the liquid. When ozone is present, on adding to the liquid the iodide-of-potassium-starch solution, the characteristic blue of iodine is developed. In the absence of ozone, no such change will take place. But on adding to the liquid a tiny drop of the transparent solution of the sulphate of iron, the iodine will at once be set free and the characteristic blue be produced.

* The metric system is used in a few of the examples which follow, in order to accustom the pupil to the mode which is adopted by all scientific men in their investigations and treatises. Any arithmetic will explain the meaning of the terms, if they are not already familiar to the scholar.

5. Let some ozone pass into a clean bottle containing a little pure mercury. Shake the whole very carefully. The metal will change so as to act like an amalgam of tin and mercury, and will form a mirror on the sides of the bottle.

6. Peroxide of hydrogen may be readily made by allowing water to drop *very slowly* through a tube containing bits of amalgamated zinc. The liquid may be tested for the peroxide by the iodide-of-potassium-starch solution in the presence of ozone, as before.

41.—1. The phosphorus will, without the aid of heat, gradually remove the O from the air, forming phosphorus anhydride (P_2O_3), which will be dissolved by the water, and in a day or two the gas which is left will be nearly pure N.

2. To show the proportion of O and of N in common air. Take a long glass tube; with a camel's-hair brush and black paint mark upon the outside the division into fifths; seal one end air-tight and fit a cork to the other. Place a bit of phosphorus in the tube near the open end and insert the cork. Heat slightly over a spirit-lamp to ignite the phosphorus, and then elevate the stoppered end so as to let the burning and melted phosphorus run slowly to the other end of the tube, combining with the O of the air as it goes. Now let the tube cool a moment, and then remove the cork underneath the surface of water, which will instantly rise and fill one-fifth of the tube. Replace the cork, and then the tube may be lifted out and shown to the class, all of whom will see that one-fifth of the air has been removed and its place occupied with H_2O, while the remaining four-fifths still appears.

42. To make the iodide of nitrogen, cover a few scales of iodine with strong aqua-ammonia. After standing for a half-hour, pour off the liquid and place the brown sediment in small portions on bits of broken earthenware to dry. They may then be carried very carefully to the class-room and exploded by a slight touch of a rod or even a feather. It can also be made by pouring aqua-ammonia upon chloride of iodine.

44.—1. For making HNO_3, a flask may be used, and the heat of the spirit-lamp will be sufficient. Take equal weights of sodium nitrate and strong sulphuric acid. Nitrate of potash will answer in place of the sodium salt. A free circulation of air is necessary. The fumes may be caught in an evolution-flask, which is kept cool by a towel frequently wet. When the retort is partially cooled, at the conclusion of the process, pour in a little warm water, to dissolve the potassium sulphate, otherwise the retort may break by the crystallization of the salt.

46.—1. A special apparatus is necessary both for preparing and inhaling nitrous oxide safely. This consists of a glass retort—as shown in the cut—a wash-bottle, and in addition a gas-bag of from twenty to fifty gallons capacity for storing the gas, and a smaller bag of from three to five gallons, with a wide, wooden mouth-piece for inhalation. It is well to pass the gas through a large wash-bottle half-full of iron sulphate solution and a second half-full of H_2O, as shown in Fig. 13, thence by a rubber tube directly into the large gas-bag. The utmost care should be taken both in preparing and administering this gas, as other oxides of nitrogen are liable to be present, especially if too high a heat is used. Before preparing the gas, pour into the bag a couple of gallons of H_2O, by standing over which it will be purified in a few hours. When about to administer the gas, let the subject grasp his nose firmly between his thumb and forefinger; then, inserting the wooden mouth-piece, be careful that he does not inhale any of the external air, but takes full, deep breaths in and out of the gas-bag. Watch the eye of the subject, and notice the influence of the gas. Great care is necessary, and no one should ever inhale the gas who is not in good health, who is troubled with a rush of blood to the head, any lung or heart disease, or is of a plethoric habit. N_2O should never be administered except when prepared and given by an experienced person.

2. Fill a small jar with the gas, and thrust into it a splinter of wood the end of which is glowing brightly; it will burst into flame.

3. Place some S in a deflagrating-spoon; kindle, and when burning briskly lower into the gas; it will burn with a pale rose-colored flame.

4. Half-fill a test-tube with gas, over water. Close the tube under water firmly with the thumb, and then agitate the water and gas together. On removing the thumb under water, a considerable rush of water into the tube will occur, as the gas is soluble in about its own volume of cold water. By this circumstance the gas is easily distinguished from O.

5. To show the effect of HNO_3 upon the metals, procure bits of tin and copper from the tinsmith. Take six wine-glasses and place them in a row upon ordinary soup-plates containing a little water. Cover each with a beaker-glass or bell-jar. In one put a strip of copper, in another a piece of silver, in another a piece of pure tin (not tinned iron), in another a strip of zinc, in another a new iron nail, in the last a bit of platinum wire or foil. Pour strong nitric acid upon each, and cover *immediately*. The copper, silver, and zinc dissolve, the latter with a violent evolution of gas. The tin is oxidized to a white powder, while the iron and platinum are unaffected. Touch the iron with a piece of copper wire, and it begins to dissolve. Put another new nail in *dilute* nitric acid, and it is rapidly dissolved.

6. Mix slowly together 1 oz. oil of vitriol and 2 oz. of the strongest nitric acid. When cold, dip paper into the mixture, and quickly wash with cold water and dry. The paper will burn with a flash like gunpowder. To avoid getting the acid on the hands, use glass tubes or rods for taking the paper out of the acid. Cotton treated in the same way becomes soluble in a mixture of alcohol and ether, and is used by photographers in making collodion. (Page 189.)

47.—1. When a jar is filled with the NO, it may be lifted out of the H_2O and inverted, when the NO_2 will pass off in blood-red clouds. If the jar be left in the cistern and one edge be lifted so as to admit a bubble of air, red fumes will fill the jar. By standing a moment, the water will absorb the red vapor. The process may be repeated several times with the remaining gas. The variation of this experiment described in the note on page 47 will be found very interesting. The change of color produced by mixing nitric oxide with any gas containing free O, often affords a convenient means of detecting small quantities of O when present in admixture with other gases, such, for instance, as coal-gas. Hence NO may be used to distinguish between O and N_2O.

2. Into a large jar inverted over water, introduce a measured quantity of NO and exactly half as much pure O. The two combine to form NO_2, which is soon dissolved in the water and disappears. This illustrates Gay Lussac's law that gases combine in simple proportions by volume. If we take 4 parts of NO to 1 part of O, we have N_2O_3, also a red gas.*

* It is *very* dangerous to breathe the red fumes (generally a mixture of NO_2 and N_2O_3) given off by the action of nitric acid upon the metals, starch, and other oxidizable bodies. The nitrous fumes produced when nitric acid is poured upon starch or sugar are passed into alcohol and sold under the name of sweet spirits of nitre. Oxalic acid (see p. 210) remains in the retort.

48.—Carefully compare the three following experiments:

1. Mix intimately 3 grams of fine iron filings in a mortar with 0.2 gram of caustic potash; introduce the mixture into a test-tube, to the mouth of which a cork and a bent tube are attached. Heat the mixture over the spirit-lamp; gas will escape, and may be collected over water in a test-tube. It burns with flame, and consists of H. At a high temperature, the Fe displaces H from the caustic potash: $Fe+2KHO = FeO+K_2O+H_2$.

2. Mix 3 grams of iron filings intimately with 0.2 gram of nitre. Heat the mixture and collect the gas as before; it will not burn, does not render lime-water milky, and is, in fact, N. The Fe has combined with the O of the nitre, forming potash and liberating N: $5Fe+2KNO_3=5FeO+K_2O+N_2$.

3. Mix 6 grams of iron filings with 0.2 gram of caustic potash and 0.2 gram of nitre, and heat the mixture in a tube. The gas which now comes off has the pungent smell of hartshorn; it is strongly alkaline, and immediately restores the blue color of reddened litmus. In the reaction which takes place, the H and the N, at the moment that each is set free, combine, and form H_3N.

4. Place a little solution of litmus, feebly reddened by the addition of a drop or two of acid, in a basin; carefully raise the flask full of ammonia gas from the gas-delivering tube; close the flask with the thumb, plunge the mouth under the solution of litmus, and withdraw the thumb; the liquid will rush rapidly into the flask, the ammonia gas will be absorbed, and the red liquid will become blue.

5. Boil a fluid-oz. of NH_4HO in a flask provided with a cork and tube, as shown in Fig. 12; the gas will come off freely. Apply a light to the jet; it will not burn readily, but a pale greenish flame will play over the top of the light. Place in a bottle of O the tube from which the gas is escaping, and then apply a light; it will now burn with a green flame.

6. "Vortex rings" formed by the fumes of chloride of ammonium. A common wooden box about three feet square has a circular hole six inches in diameter cut in its front face, while the back of the box is removed and its place supplied by a piece of tightly-stretched canvas or linen cloth. Let the vapors of ammonia and hydrochloric acid pass constantly into the box through small holes at the side. When the box is filled with the white fumes of the chloride of ammonium, a sharp but gentle tap on the canvas back will drive out, through the hole in front, a *beautiful ring of smoke* that will traverse a large room and will have force enough to extinguish a lighted candle.

50.—1. For preparing H the apparatus shown in Fig. 13 is very convenient. The wash-bottle, *d*, is necessary only when it is desired to purify the gas for inhaling. A common junk-bottle, fitted with a cork and a glass tube, will answer for all ordinary experiments, but a "hydrogen generator," as described below, is much more satisfactory. The Zn for making H should be granulated.* Water may be poured into the flask until the lower end of the funnel is covered, before adding the acid. The flow of gas may be regulated by additions of acid, as may be wanted. One part of acid to 10 or 12 parts of water will liberate the gas rapidly. If too much H_2SO_4 be added the liquid is apt to froth over.

A constant hydrogen generator can be readily made by taking two bottles with tubulature near the bottom, such as are sold by druggists for the "nasal douche,"

* This is easily done by melting the Zn in an iron ladle, and pouring the metal slowly from a little height into a basin of water.

and connecting them with a strong rubber tube. One is fitted with a cork and delivery tube provided with a stop-cock. In this bottle is placed a layer of pebbles or broken glass, and upon this a quantity of zinc scraps. In the other bottle is dilute H_2SO_4. On opening the stop-cock the acid comes in contact with the zinc and H is evolved. As soon as the stop-cock is closed, the pressure of the gas drives the acid back into the other bottle. The same kind of apparatus may be employed for generating CO_2 or H_2S.

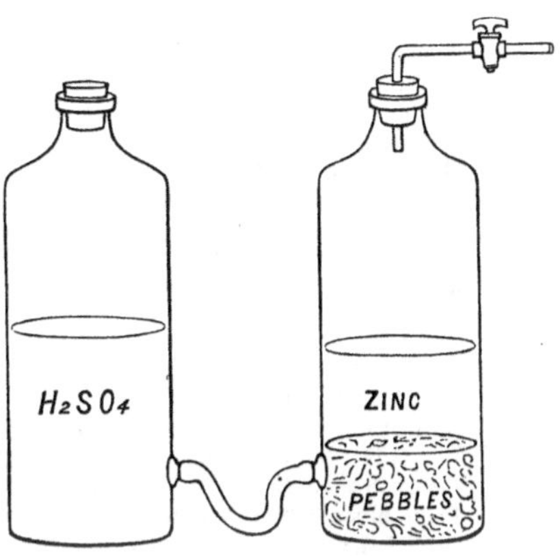

A hydrogen generator, similar in principle to the Döbereiner's lamp (Fig. 18), can be made by cutting off the bottom of a tall and narrow bottle, filling it with zinc scraps and closing at the lower end with a perforated rubber cork, and at the upper end with a perforated cork carrying a brass or glass tube and stop-cock. Place it upright in a jar of dilute sulphuric acid.

In experimenting with H, great care must be used not to ignite the jet of gas until all the common air has passed out of the flask; otherwise a severe explosion will ensue.* It is a safe precaution to test the gas by passing it in bubbles up through H_2O, and igniting them at the surface; the force of the combustion will indicate if there be any danger. H must not be kept in bags for any great length of time, as the air will gradually force itself in, and the gas will partly pass out by the law of diffusion, thus forming a mixture which it is dangerous to ignite.

2. The gases may be mixed in the following manner: Fit a good cork into the neck of a large jar, and pass through it a tube 5 centim. long. Bind a short piece of rubber tubing firmly to the tube, and close this elastic tube with a small screw-vise.† Fill the jar with water over the pneumatic tub. Fill a small jar which will hold about half a litre with O, and transfer it, as shown in Fig. 17, to the large jar. Fill the same jar with H, and transfer it to the large jar. Repeat the operation with the H, so as to obtain in the larger jar a mixture of half a litre of O and 1 litre of H. Having previously softened a thin bladder by soaking it in water, tie into the neck of it a glass tube 5 centim. long; then adjust to the projecting portion a piece of rubber tubing provided with another nipper-tap. Press the air out of the bladder; connect by means of a short piece of glass tubing the two pieces of rubber tube; depress the jar in the pneumatic tub, and then open

* Always wrap a cloth around the H generator when you ignite the gas, as an explosion may take place at any time.

† Small vises, or "nipper-taps," as they are called, are sold for this purpose. They are cheaper than stop-cocks, and answer every purpose. In lieu of these, common spring clothes-pins may be used.

each nipper-tap. The gas will now pass into the bladder; if it does not, press the jar deeper into the water; close both nipper-taps, and remove the bladder. Now place the end of the tube attached to the bladder under some soap-suds, and force out the mixed gases by squeezing the bladder so as to make a lather. *Carefully remove the bladder to a distance*, and then apply a light to the froth of soap-suds. A loud explosion will immediately follow.—A clay tobacco-pipe may be attached to the gas-bag by means of a bit of rubber tubing. Dip the pipe-bowl into the soap-suds, and lifting it out, blow a bubble with the mixed gases, and then detach it by a quick motion. When the gas-bag is removed, ignite the bubble, which will explode sharply. If bubbles be blown with H alone, they will rapidly rise, and if out of doors, will float to a great distance.*

3. H is the lightest known substance. Fill two bottles with the gas, suspend one inverted from the ring of the retort holder and place the other right-side up on the table. In a few minutes the upright cylinder will be found to contain little or no H, while the inverted bottle is nearly full.

4. Invert an empty beaker glass or paper box, and suspend it from the end of the scale beam (Fig. 26). Balance it carefully, then fill it with H gas. It will be found much lighter than before.

5. Take a small porous cup, such as is used for galvanic batteries. Fit a cork to it and pass a long glass tube through the cork. Cover the cork with plaster of Paris. Place it upright with the lower end of the tube dipping into a colored liquid ($CuSO_4 + NH_4HO$). Hold a jar of H over the porous cup. The H enters the cup, driving the air out of the lower end of the tube. Remove the jar and the liquid will rise several inches in the tube. (See Physics, p. 50.)

6. A substitute for spongy platinum in the experiments with hydrogen gas. Make a cylinder of pumice stone ⅜ of an inch in diameter. With a fine saw cut it into disks about one-twentieth of an inch thick. Soak these for some time in a strong solution of bichloride of platinum in alcohol, and then as long in an alcoholic solution of sal-ammoniac. After being once thoroughly ignited these disks will inflame a jet of hydrogen.

60.—1. The analysis of water † can be readily performed as follows: Take a wide bottle (the height is unimportant) and cut it off about 2½ inches below the neck, by making a scratch with a three-cornered file and then applying near the scratch a very hot piece of wire or glass, or a small blowpipe flame. The crack will follow the flame slowly around the bottle. Take two strips of platinum foil, and put one on each side of a well-fitting cork, so that one end extends into the bottle, the other outside of the neck. Support the apparatus inverted upon a ring of the retort stand, and fill nearly full of water acidulated with H_2SO_4. Connect one strip of Pt with each pole of a galvanic battery. Bubbles of gas at once appear and can be collected in inverted test-tubes and tested. Instead of a bottle neck, a broken funnel may be employed. (See Physics, p. 237.)

2. The synthesis of water may be shown, and also its composition by weight, by passing dry H over dry CuO. The CuO is placed in a bulb-tube of hard glass *C* and weighed. The chloride of calcium tube *D* is also weighed. The H

* If one has a large rubber gas-bag with stop-cock and rubber tubing, and a glass receiver fitted with a stop-cock on top, these may be attached and the gases measured in the receiver and then passed directly into the bag. Such apparatus, though convenient, is not necessary to illustrate the properties of the gases.

† The so-called *Water Gas* is made by passing H_2O over red-hot coal. It is chiefly H + CO.

generated in *A* is dried by $CaCl_2$ at *B*, and passes over the CuO at *C*. When the air has *all been expelled*, heat the CuO until it has a bright red color. Allow the H to pass through until the tube *C* is cold. Take the apparatus apart and weigh the tube *C*; it will have decreased in weight by a quantity equal to that of the O expelled. Weigh the tube *D*. It has increased by a quantity equal to that of the water formed. If *C* has lost 16 grams in weight, *D* will have increased

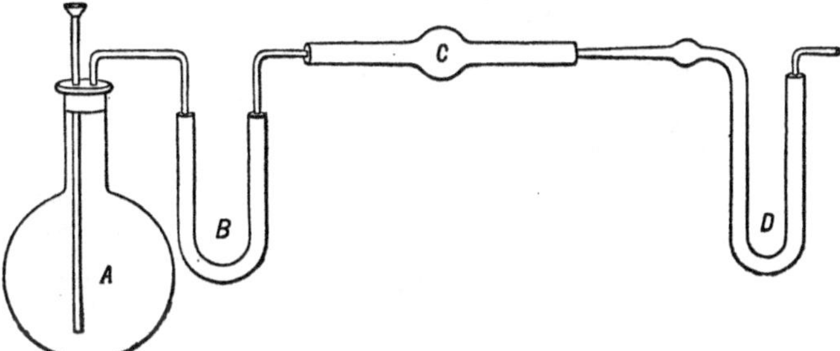

18 grams in weight, showing that 16 parts of O will form 18 parts of water, by taking up two parts of H.

3. Burning H in O. Attach a $CaCl_2$ drying tube to a hydrogen generator, and to this a glass tube bent twice at right angles and then turned up at the end, as shown in the figure. Take a small piece of Pt foil and roll it around a darning needle so as to form a small tube. Soften the end of the glass tube and slip this little tube into it while hot, then hold them in the flame until the glass settles down against the Pt on all sides. When the air has all been expelled from the generator, throw a towel over it and ignite the H, then bring it into a broad jar of O. The heat is very intense, hence the need of a Pt tip.

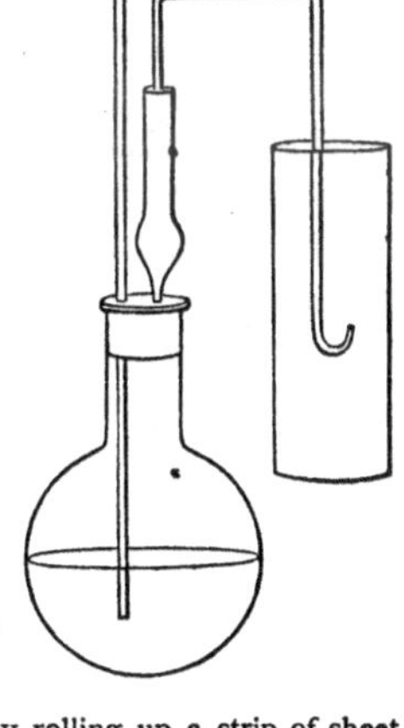

4. Burning O in H. To show the reverse experiment of burning O in H, or illuminating gas, take a large lamp chimney and fit a cork in each end. In the upper cork insert a glass tube drawn out to a jet *D*. In the lower end insert a bent glass tube *A* and a metal tube *C* made by rolling up a strip of sheet iron or brass. Fit a cork to the metal tube and pass a glass tube with a Pt tip on it through this cork *B*. Attach the tube *A* to an H generator, and when the whole appara-

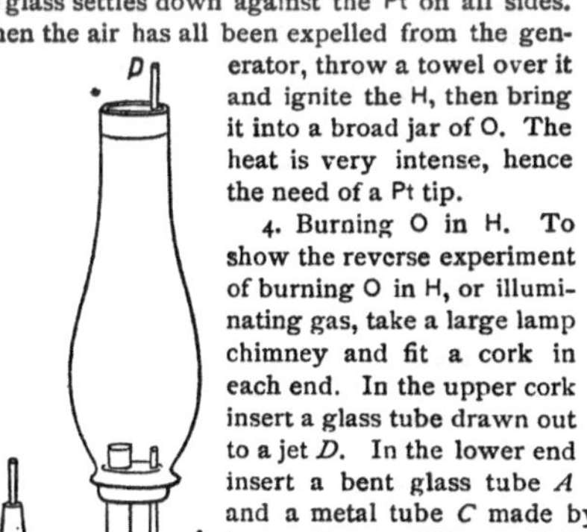

tus is *full* of H, ignite it at *D* and *C*. Connect *B* with a gasometer or bladder of O, then quickly insert the cork into the opening *C*, so as to extinguish the flame there. Let the O pass in *slowly*. The O will be seen to burn in an atmosphere of H. Cut off the supply of O and extinguish all flames before removing the supply of H, to avoid an explosion.

59.—1. Grind in a mortar 50 or 60 grams of sodium sulphate with about twice its weight of water at 15° C. The water will dissolve a considerable portion, but not the whole of the salt. Pour this saturated solution into a flask, and warm it gently; it will now dissolve 50 grams more of the salt without difficulty. When all the salt is dissolved, cork the flask, place it on the table and allow it to cool slowly. Be careful not to shake it when cold. On removing the cork the whole becomes a mass of crystals, which can be seen to shoot through the liquid in every direction. Pour off the liquid, and dry the crystals by pressing them between a few folds of blotting-paper. When they appear to be dry, put a small quantity of the crystals into a test-tube, and apply a gentle heat; the salt will liquefy, and on continuing to apply the heat a large quantity of water will be driven off, and a dry, white powder will be left in the tube.

2. Take some of the fresh crystals of sodium sulphate; let them lie exposed on a piece of blotting-paper for two or three days. They will gradually lose their water and crumble down, or *effloresce* into a white powder.* Dissolve 150 parts, by weight, of hyposulphite of soda † in 15 parts boiling water, and gently pour it into a tall test-glass so as to half fill it, keeping the solution warm by placing the glass in hot water. Dissolve 100 parts, by weight, of sodium acetate in 15 parts hot water, and carefully pour it in the same glass; the latter will form an overlying layer on the surface of the former, and will not mix with it. When cool there will be two supersaturated solutions. If a crystal of sodium hyposulphite be attached to a thread and carefully passed into the glass, it will traverse the acetate solution without disturbing it, but, on reaching the hyposulphite solution, will cause the latter to crystallize instantaneously in large rhomboidal prisms. (Compare p. 133.)

4. Take two four-ounce bottles, and put in each a teaspoonful of white sugar. Fill one bottle nearly full of pure rain or distilled water, the other with impure water. Cork them, and let them stand a few days; in one bottle will be seen a fungoid growth, resembling fuzz or lint; the water in the other bottle remains clear. (Note, p. 192.) The sporules or germs that fall into the water find suitable nutriment for their development in one case, but not in the other. *Any well or spring water in which this change takes place is unfit to drink.* (P. 60, note.)

5. Select a thin, porcelain dish which will hold 60 or 80 cub. cm.; place it in one pan of the balance, and trim a piece of lead until, when placed in the other scale-pan, it will counterpoise the dish. Measure a quarter of a litre of spring-water, and pour some of it into the weighed dish; place it over a very small gas-flame, so as to evaporate the H_2O gently, without allowing it to boil; add the rest of the H_2O from time to time until it has completely evaporated. Dry the salts thus obtained, and weigh what is left as accurately as you can. By multiplying this quantity by 4, you will obtain the amount of soluble solid substances per litre which that particular specimen of water contained. This is the basis of the plan which, with many additional precautions, is adopted for determining the

* Common washing soda, $Na_2CO_3 + 10H_2O$, will do the same.

† To be had of any photographer, under the name of "hypo."

quantity of salts in the process of analyzing waters to be used for drinking or manufacturing purposes.

66.—1. Small paste-diamonds may be obtained of a jeweller, to illustrate the forms of cutting the diamond.

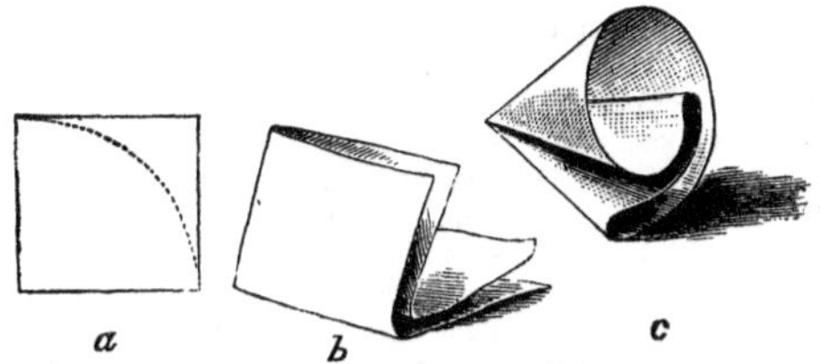

69.—1. Place a filtering-paper in the glass funnel,* and in it two ounces of bone-black or finely-powdered charcoal. Filter through it water colored with ink, litmus, or any other impurities. In pouring the liquid into the filter, hold a glass rod against the edge of the pouring vessel, so as to direct the stream into the funnel. The funnel may be placed in the nozzle of a bottle, but must not fit closely. A bit of wood or a thread inserted between the stem of the funnel and the nozzle will leave an opening sufficient for the egress of the air.

2. Slip a piece of freshly-burned charcoal under the edge of a long tube previously filled with dry ammonia gas,† and standing over Hg. The charcoal will quickly absorb the H_3N; the whole of the gas, if pure, will disappear, and the Hg will fill the tube.

3. Weigh a piece of freshly-burned charcoal as soon as it is cold; leave it exposed to the air for twenty-four hours, and weigh it again; it will be found to be heavier. Place the charcoal in a glass tube, and heat it over a lamp; moisture will be driven off, and will become condensed on the cold sides of the tube.

4. Shake up with a little powdered charcoal some stagnant water which has been kept till it smells offensively. In an hour it will have lost all its disagreeable odor.

5. Mix in a mortar twenty grams of litharge with forty grams of NaCl and one gram of powdered charcoal; cover with a little more salt, and place the mixture in a small, clay crucible; heat it to bright redness in the fire. When the mixture is melted, take the crucible out of the fire and let it cool. When quite cold, break the crucible, and a bead of Pb will be found at the bottom, under the melted salt, the C having taken the O from the PbO.

6. Break the head off a match, and warm the match-stick in the flame. At the same time hold one end of a large crystal of sal-soda in the flame until it melts. Rub the stick with the melted soda until it is well covered; then hold it in the flame until the protected stick is completely charred. Mix a very little PbO with some melted soda and put on the end of the stick, and place it for a

* In order to prepare this filter, fold a square of paper, as shown in the figure above, first into half, and then again into a quarter of its first size (*b*); cut off the edges in the direction of the dotted line shown in the left-hand figure (*a*), open out the folded paper (*c*), and drop it into a funnel a little larger than the paper cone.

† The gas may be dried by passing it through a tube filled with pieces of calcium chloride (see Fig. 16), obtained in making CO_2. (See page 74.)

short time in the centre of the flame. On taking it out, minute beads of lead can be seen on the stick. Pulverize the end of the stick in a small mortar, and pour on some water. The pieces of charcoal can be washed away, leaving little bright spangles of lead, which flatten under the pestle.

7. Repeat the experiment with a small quantity of copper oxide. The metal will require a stronger heat, but may be reduced in like manner. If the little bead be placed between two folds of paper, it may be flattened with the hammer, and will show the red color of copper.

8. Select a small stick of charcoal, and with the point of a knife make a small cavity of the size of a split pea near one end. Put a little white lead in the cavity, and heat it strongly before the blowpipe in the reducing flame. A little bead of lead will easily be obtained, surrounded by a border of yellow lead oxide. The lead will flatten under the hammer.

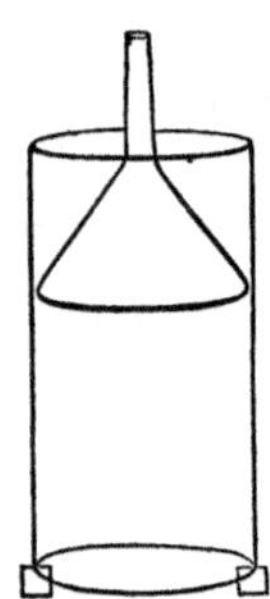

9. Take a glass cylinder open at both ends, and suspend near the top an inverted funnel which fits nicely into the cylinder. Place a turpentine lamp below it so that the smoke passes up into the funnel. When a considerable quantity of lampblack has formed on the sides of the cylinder and funnel, extinguish the flame and remove the lamp. Then slowly lower the funnel, the edge of which will scrape the lampblack from the sides of the vessel, in precisely the same manner that it is done by manufacturers in making lampblack for the market.

74-6.—1. Break some marble into small bits; place them carefully in the evolution-flask, and, inserting the cork and tube, pour in HCl slowly. The gas, on account of its weight, may be passed directly into a bottle or jar.

2. Lower a lighted candle into a jar of the gas, or, placing the candle in an empty jar, pour the gas into the jar, as if it were water. Test the acid with blue litmus-paper moistened.

3. In a pint of water place a piece of lime as large as an egg; let it stand over night; pour off the clear liquid; it is lime-water. Place a little in a tumbler and breathe through it by means of a tube, or pass a current of CO_2 from the evolution-flask until the liquid, at first milky, clears.

4. Breathe through a tube into an empty bottle. Lower into it a lighted candle—it will be immediately extinguished. Pour in some lime-water, shake thoroughly, and it will become milky.

5. Twist a wire around the neck of a small, wide-mouthed vial, to answer as a bucket. Lower it by the wire into a jar of CO_2, our ideal well foul with the gas. Raise it again, and test for the CO_2 by means of a lighted match. The bucket will be found full of the gas.

6. Balance a large paper bag or box on a delicate pair of scales, or in any simple manner one's ingenuity may suggest. Empty into the box a large jar of CO_2, and the box will quickly descend.

7. Arrange little wax-tapers in a wooden or pasteboard trough, as on page 75. Light them, and then pour in at the top a bottle of carbonic acid gas. If the proper slant is given to the trough, all the candles will be extinguished.

8. To show the formation of carbonic acid in the lungs, repeat Faraday's favorite experiment. Fit an open-mouthed receiver with a cork, through which passes a small bit of glass tubing. The receiver is then placed in a basin of water.

Expelling the air from his lungs, the experimenter inhales the air in the receiver through the tube. The water in the basin rises in the receiver, and shows how fast and when the air is exhausted. Breathing back the air from the lungs into the receiver again, the water is expelled, and the lighted taper will test the presence of the carbonic acid. Before testing, the air may be breathed back and forward two or three times, until it becomes unpleasant. The rise and fall of the water in the jar is a pleasant and instructive addition to the experiment.

80.—1. Dry some potassium ferrocyanide, K_4FeCy_6, $3H_2O$ (prussiate of potash), till it crumbles down to a white powder. Mix 5 grams of this with 50 c. c. of oil of vitriol in a flask; adjust a cork and a wide, bent tube to the mouth of the flask, and heat the mixture. The CO will come off very quickly, and will burn with a blue flame. Take care *not* to inhale it.

2. Place a few crystals of oxalic acid ($C_2H_2O_4$) in a test-tube, pour on enough oil of vitriol to cover them, and then heat gently. Both CO and CO_2 will be evolved. On applying a flame to the open end of the tube, the CO will take fire and burn with a blue flame. To separate the two gases, pass them through a solution of KHO, which will remove the CO_2, and the CO can then be collected over water.

82.—1. Introduce into a retort which will hold a litre, 15 c. c. of alcohol and 60 c. c. of oil of vitriol. Heat the mixture, and collect the gas over water; continue the experiment until the mass blackens and swells up considerably. The product consists at first chiefly of olefiant gas, mixed with ether-vapor; but towards the end it becomes mingled with SO_2. Pass it through a solution of potash, using a wash-bottle as shown in Fig. 13, and then collect in the gas-bag. Fit a piece of glass tubing, drawn to a fine point at one end, to the stop-cock of the gas-bag, by means of a bit of the rubber tubing. On turning the stop-cock and forcing out the gas, it may be ignited, when it will burn with a clear white light.

2. Mix C_2H_4 with twice its bulk of O and explode in soap-bubbles. It produces a greater noise even than the "mixed gases." Great care must be taken not to let the light approach the gas-bag containing the mixture.

3. At the close of the first experiment, perform the one described in the note on page 89. A small piece of wire-gauze, four to six inches square, for this purpose can be purchased of any tinsmith. If you do not force the gas out too rapidly, you will be able to burn it on either side of the gauze at pleasure.

4. Place on top of the gauze a piece of camphor-gum. Ignite it, and the flame will not pass through to the lower side. Then ignite on the lower side, and extinguish the flame on the upper side.

83. Get a bit of bituminous coal about the size of a walnut. Pound it small, almost into dust. Fill an ordinary tobacco-pipe (one with a long stem is preferable), *nearly* full of the pounded coal, packing it down closely with your thumb. On the top press a disk of metal or a copper coin, and cover with a layer of plaster of Paris or some tough clay, reduced to the consistency of putty by being tempered with a little water. Heat the bowl of the pipe strongly, and a combustible gas will come out of the stem, which should now be held in a nearly vertical position. When no more gas is given off, and the jet of flame goes out, remove the clay or plaster covering. The residue in the bowl is coke.

2. Place some bits of pine wood in a glass retort provided with a perforated cork and delivery-tube. Connect this with an empty wash-bottle. On heating the retort, gas is given off, tar collects in the wash-bottle (see p. 205), and char-

coal remains in the retort. The charcoal used for making gunpowder is prepared in this way in large iron cylinders.

84.—1. Fit a cork to a small test-tube. Take out the cork, and pass through it a bit of glass tubing drawn to a fine point at one end, so as to act as a gas-burner. Place in the tube fifteen or twenty grains of mercury cyanide; replace the cork, and heat over a spirit-lamp. The test-tube may be supported by a strip of thick paper twisted around it at the top. Move the tube to and fro through the flame at first, until it becomes fully heated; hold the tube inclined and not perpendicular, letting the flame strike the side rather than the bottom. When the gas begins to come off, it may be ignited.

2. To show the formation of potassium cyanide from a nitrogenous body. Take a thin piece of freshly-cut K, lay on it a bit of the nitrogenous substance, and cover with another thin piece of K. Press all tightly together; drop the mass into a perfectly dry test-tube, and heat carefully until it melts and a flash of light is seen. When cold, break the tube, throw the fused mass into a clean test-tube, and make the test for KCy with $FeSO_4$ and Fe_2Cl_6, or as given below.

3. To test for the presence of potassium cyanide. Add to a solution of the suspected substance a few drops of solution of sulphate of iron and a slight excess of potash. Shake the precipitate a few moments with air in the tube, and add an excess of hydrochloric acid, when a blue precipitate, or a decided blue or green color pervading the liquid, will indicate the presence of a cyanide. Prussian blue is produced by this process.

4. To show that there is no free hydrocyanic acid in the kernels of peach, cherry, and plum stones, or bitter almonds, but that it is formed on heating the same with water. Soak a long strip of Swedish filter-paper in a tincture of gum guaiacum (1 to 20), and dry. Next pass through a solution of sulphate of copper diluted 2000 times, when the paper will not be changed at all in color. Put a few freshly-pounded bitter almonds in a two-liter flask with water. On suspending in it the strip of test-paper above described, the paper will remain white; but on pouring into the flask a single crushed bitter almond that has been *warmed* with water, the test-paper will at once be colored blue by the hydrocyanic acid generated in the flask, without bringing the paper in contact with the liquid.

85.—Take an ordinary argand lamp with an opening at the bottom to admit air at the centre of the flame. Raise the wick till it smokes. Apply the finger to the opening, and the flame becomes more dull and smoky. Remove the finger, and the flame brightens again. Now admit a jet of O at the opening, and the flame will whiten almost to the intensity of a lime-light. Great caution must be used, unless an oil-lamp be employed. The bright, clear flame shooting instantly through the cloud of smoke is very striking.

86.—The structure of a flame is easily shown by taking a piece of white card-board about 3 inches square, and placing one edge against the wick of a lighted candle below the flame, and then bringing it quickly to a vertical position *in* the flame for an instant. On removing the card, there is burned on it a picture of the flame similar to Fig. 31.

90.—Take the beak from a broken retort, draw out one end so that it will enter the air-hole *b* (Fig. 37) of a Bunsen burner. Place a turpentine lamp under the open end of the beak, held at an angle of 45°. The smoke enters the burner, and the non-luminous lamp becomes luminous, because the carbon is all burned.

91.—1. The compound blow-pipe with gasometers, as shown in Fig. 38, is a serviceable apparatus. If gas-bags are used, the one for H should be twice the size of the one for O. A board should be laid on each bag, upon which weights may be placed, when ready for use, so as to force out the gas steadily. Turn the stop-cock so that the H will pass out twice as fast as the O. Always ignite the H first, and then turn on the O slowly until the best effect is produced. If gasometers are used, press the inner receivers down to the bottom, and then pour in water until it reaches nearly the top. The rubber pipes may then be attached to the hydrogen or oxygen apparatus, and the gases passed directly into the gasometer. Proper pressure is produced, when the jet is to be ignited, by unloosing the strings from the inner receivers, and thus taking off the "lift" of the weights which equipoise them. Additional pressure is secured by bearing down upon the receivers. All the metals burn in the blow-pipe flame with their characteristic colors. Narrow slips should be prepared for this purpose. A mirror, and a cup for holding the chalk, are necessary to show the lime-light. A piece of hard chalk or lime, whittled to about the size of a pencil, may be held in the flame to illustrate the principle.

99.—1. To a small gas-jar fit a good cork, through which pass a test-tube as shown in Fig. 42. Place the jar in a large beaker-glass or open-mouthed bottle, filled with spring water, which has been mixed with a fourth of its bulk of a solution of carbonic acid in water. Fill the tube with water, and place it in the neck of the jar, having introduced a few sprigs of mint or the leafy branches of any succulent plant; then expose for an hour or two in direct sunshine. Bubbles of gas will be seen studding the leaves; and on shaking the jar they will become detached, and will rise into the test-tube. After a time the cork and tube may be withdrawn, keeping the mouth of the tube beneath the surface of the water; then close it with the thumb, turn the tube mouth upwards, and test the gas with a glowing splinter. The wood will burst into a blaze, showing that the gas consists mainly of O.

102.—1. Put in the flask two ounces of NaCl and an ounce and a half of MnO_2. Pour on enough water to reduce the mixture to a thin liquid. Shake the flask until the whole interior is moistened. Insert the cork and delivery-tube; the middle bottle (B), shown in Fig. 43, is not necessary. Fill the pneumatic tub * with strong brine or *warm water*, *using as small a quantity as possible*, since water absorbs the gas. Pour in an ounce of H_2SO_4 through the funnel-tube (F), or directly at the nozzle, by removing the ground stopper, if a kind of flask be used which has one. The gas will come off at once, even before the heat is applied. Collect the gas in bottles and use directly, if convenient, otherwise put corks in them and rub the nozzles well with tallow.† Pass the gas through a tumbler of cold water; this will form chlorine-water, which should be bottled and kept in a dark place, or wrapped in black or yellow paper.‡

2. Plunge a lighted taper into the gas: it will burn feebly, with a red, smoky flame.

3. Place a piece of dry phosphorus in a copper deflagrating-spoon; introduce it into a bottle of Cl: the phosphorus will take fire, and burn with a pale greenish flame, while suffocating fumes of phosphoric chloride (PCl_5) are formed.

* If this be large, use a tin pan in its place, and have a pail of warm water for filling the bottles.

† A better way is to fill the wash bottle B with H_2SO_4 and collect in *dry* glass stoppered bottles by displacement.

‡ If the water is cooled to 5° C. while the Cl is passed into it, the whole becomes a mass of crystals, hydrate of chlorine.

4. Dip a strip of blotting-paper into oil of turpentine; plunge it into a jar of Cl: it will immediately burst into flame, while a dense black smoke is given off.

5. Powder some metallic Sb finely in a mortar, and sprinkle into a jar of Cl: it will take fire as it falls, giving out fumes of antimony chloride ($SbCl_3$), which are very irritating.

6. Pour a little boiling water upon some chips of logwood, so as to obtain a deep red liquid: add some of the solution of Cl, and the red color will be discharged.

7. Wrap a soda-water bottle in a towel; fill it with water, and invert it in the pneumatic tub. Introduce a glass funnel into the neck, and having filled a jar of 100 c. c. capacity with Cl, pass the gas into the bottle. Fill the same jar with H, and empty into the same bottle; withdraw the funnel, close the neck with the palm of the hand, lift the bottle out of the water-bath, give it a shake to mix the gases, and apply a light. A sharp explosion will immediately follow, and gaseous HCl be formed. Equal measures of H and Cl unite in this way, and the gas produced occupies the same bulk that its components did when separate. Sunlight will also cause the explosion of a mixture of Cl and H.

8. Fill a large jar with equal volumes of chlorine and olefiant gas (p. 82). Leave it standing for a while in diffused daylight; oil drops will soon be seen standing on the sides of the vessel. It combines directly with the chlorine to form $C_2H_4Cl_2$, formerly called "The oil of the Dutch chemists." To this fact olefiant gas is indebted for its name, "oil builder."

9. Mix together 2 volumes of Cl and one of C_2H_4 in a wide-mouthed jar, remove the cover and ignite. The Cl combines with all the H, setting free all the C, which ascends as a thick black smoke.

10. Burn H in Cl, as shown in experiment of burning H in O.

11. To show the manner in which Cl acts as an oxidizing agent. To a solution of $MnSO_4$ add lime water and Cl water. It turns black from the formation of MnO_2.

12. Print the word PROTEUS on a large card, first with the iodide-of-potassium-starch solution (note, p. 107), and second with a solution of indigo. The former will be white and almost invisible, the latter blue. Then paint the words (using a camel's hair brush) with a solution of chlorine. The first line will turn blue and the second white, thus just reversing their color.

105.—1. Melt 200 or 300 grams of NaCl in a clay crucible at a good red heat, and, when melted, pour out the salt upon a dry stone slab. On cooling, break the mass into pieces of the size of a pea, and preserve them in a dry bottle. Introduce 50 grams of the chloride, and about twice its weight of H_2SO_4, into a flask provided with a cork and bent tube. Cl gas comes off, even in the cold, but it is extracted still more abundantly when heated. Collect the gas in dry bottles by displacement. It may easily be ascertained when the bottle is full, as a lighted taper will be extinguished if introduced.

2. Fill a long-necked flask with the gas by displacement, close the neck with a perforated cork in which is a glass tube drawn out at the end, and immerse it in a basin containing infusion of litmus; the blue liquid will rush into the flask, forming a fountain, and will become red.

3. Fill a dry bottle by displacement with Cl gas, and close the mouth with a glass plate. Withdraw the stopper from a bottle of the same size containing ammoniacal gas; invert the jar of Cl over the one containing the H_3N, and

remove the glass plate. The two invisible gases will suddenly combine, a dense white cloud will be formed, and a solid salt produced.

4. Dilute a little HCl with 6 or 8 times its bulk of water, and add caustic soda cautiously, until the liquid is neutral, and neither reddens blue litmus nor restores the blue to red litmus-paper. Pour the liquid into a basin, and evaporate it slowly; crystals of NaCl will be deposited in cubes.

5. Boil HCl in a test-tube with fragments of gold leaf: they will not be dissolved. Now add a drop or two of HNO_3: a yellow solution of gold chloride ($AuCl_3$) will be quickly formed.

6. Fill a test-tube nearly full of pure rain or snow water, and add a drop or two of the nitrate of silver solution. A drop of HCl will cause a cloudy, white precipitate.*

106.—1. Grind in the mortar 3 or 4 grams of fluor spar, and mix with an equal weight of powdered glass. Introduce it into a flask previously fitted with a sound cork and a tube, as in the figure. Pour upon the mixture 30 grams of H_2SO_4, insert the cork and tube, and apply a gentle heat: a densely fuming gas is disengaged, consisting of silicic fluoride (SiF_4). This gas must not be inhaled, as it is very irritating. Pass it into a glass of H_2O, having sufficient Hg at the bottom to cover the mouth of the delivery-tube. Each bubble of gas as it rises is coated with a white film of hydrated silica, while the H_2O forms a solution of hydrofluosilicic acid ($2HF,SiF_4$). The deposit of silica would clog the tube if it were not for the Hg; hence the tube must be kept dry, which is best accomplished by placing it in position in the Hg, then pouring water carefully into the glass on the Hg. Filter the solution and preserve it as a test for Ba and K.

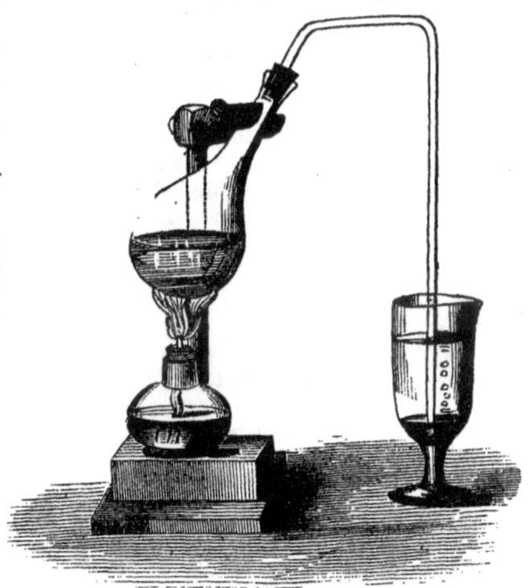

106. Fill a long vial half-full of water colored with bromine. Add ether to fill the vial. The ether will float on the water and the vial will contain on top a colorless and at the bottom a red liquid. Shake the vial thoroughly, and, if the proportion of the ingredients be properly calculated, the ether will dissolve the bromine, and, on replacing the vial and giving time to settle, there will be a red liquid on top and a colorless one at the bottom. This illustrates the varying power of solution possessed by different liquids.

* Well water which gives a considerable precipitate with $AgNO_3$, is usually unfit to drink owing to sewage contamination. (See pp. 256, 60.)

107.—1. Fill three test-tubes nearly full of soft water.* Pour in one a few drops of a solution of mercuric chloride; into the second, of sugar of lead; into the third, of mercury subnitrate. Add to each of these a few drops of a solution of potassium iodide. The first especially will produce a brilliant color, mercury iodide; the rapid change from yellow to red is very marked. On continuing to add the potassium iodide, the red precipitate will be dissolved and disappear.

2. Make an additional quantity of mercury iodide. Let it settle. Pour off the liquid, and then spread the sediment on a piece of heavy card-board, making a red spot as large as a silver dollar. Dry it carefully. Then heat very strongly, when it will turn yellow. Rub over the yellow spot the point of a knife several times, bearing on very firmly, until a red mark can be seen. Lay away the paper for a day or two, and the red color will spread over the whole spot. As a variation of this experiment to show how using different quantities of a substance may produce different results. Fill a glass half-full of the scarlet mercury iodide; also a larger glass with the potassium iodide solution. Pour the thick scarlet liquid into the transparent one, and the color disappears as if by magic, and a glass of apparently limpid water remains. Yet this second liquid which seems to swallow up the first was a part of the same which, when added to the mercury solution, originally produced the color.

108.—1. Bend the end of a piece of thin platinum wire, 8 or 10 cm. long, into a small hook; heat the wire to redness, and instantly touch with the wire a crystal of borax as large as a split pea; it will adhere to the wire. Then introduce the wire and crystal into the flame of a spirit-lamp. The borax will swell up, become opaque and white, and will then melt into a clear, glassy bead.

2. Touch the bead just made, with a wire moistened with a solution of cobalt nitrate. Then melt the borax again in the flame. A beautiful blue bead is obtained, which is almost opaque if the quantity of cobalt be considerable. If a scarcely visible fragment of manganese oxide be used, a violet bead is formed.

3. Dissolve a few crystals of the boracic acid in a small dish with a teaspoonful of alcohol. Set fire to the spirit: it burns with a green flame, which is a good test for boracic acid. A similar green flame is obtained if a crystal of borax be moistened with glycerine and kindled as before.

110.—1. Grind a little glass to a fine powder in a mortar; place it on a piece of moistened red litmus-paper; sufficient alkali will be dissolved by the water to tinge the paper.

2. Mix a little fine sand with KNO_3 and Na_2CO_3, and heat strongly on a strip of Pt foil for 5 minutes. When cold it will dissolve in H_2O.

3. Take some of the silica obtained in the experiment of making $2HF,SiF_4$, and put it in a strong solution of KHO and boil. It will dissolve.†

4. Take four glass cylinders 5 in. high, and pour into each about 1 oz. of ordinary water glass and 4 oz. of water. Drop into one a few crystals of iron sulphate, into another some crystals of blue vitriol, into the third white vitriol, into the fourth a crystal of each. Let them stand quietly for 24 hours. In the first, green fibers will be seen resembling very closely a growing plant; blue and white ones will appear in the others. If closely corked they can be kept for weeks.

* Melted snow, or very clear rain-water, will answer the place of distilled water in making solutions, etc., for experiments.

† The infusorial silica, sold under the name of electro-silicon, will dissolve in KHO in the same manner. A basic silicate known as "water glass," or "soluble glass," is prepared in this way.

5. Pour 1 oz. of water-glass into a capsule and pour *on it* half as much H_2SO_4, taking care that the two do not mix. Pour immediately, but slowly, into a second capsule; the silica separates in long tubes resembling stalactites.

113.—1. Melt a quantity of S, either the flowers or brimstone, in a test-tube. It is at first thick and dark-colored, but after continued heating regains its fluidity. Warm the side of the tube all the way to the top, and pour the liquid into water. It will form an elastic gum, which can be moulded into any desired form.

2. Heat a piece of brimstone in a test-tube. After a little the S will sublime and collect in the upper part of the tube as flowers of sulphur.

3. Fill a cup with brimstone and melt it with a gentle heat. Set it aside to cool. When a crust has formed on top, break it and pour out the liquid contents. If the cup be broken when cold, the bottom will be found covered with long crystals of S.

4. Pulverize some brimstone and put in a test-tube and cover with CS_2. Put the tube in warm water until the S dissolves, then cork it tightly, or, better, place it on a watch-glass under a bell-jar, and let it stand until crystals separate. They have a different form from those obtained from fusion. The more slowly they separate the larger they will be.

5. Put a piece of S in a test-tube and above it some copper turnings or scraps. Heat the sulphur until it boils. The Cu burns brightly in the vapor of S. Fe and Pb will also burn in S vapor.

114.—1. SO_2 is conveniently prepared by heating H_2SO_4 and Cu (Fig. 43), drying with H_2SO_4, and collecting by displacement, as it is quite soluble in water.

117.—1. Pour a little strong sulphuric acid into a test-tube. Place a splinter of wood in it: the wood will be blackened in a few minutes. Pour 1 c. c. of strong H_2SO_4 into a tube containing 3 or 4 c. c. of water: considerable heat will be felt to attend the mixture.* Take a little of this diluted acid, and with a feather dipped into it trace a few letters upon writing-paper. Hold the paper near the fire: the water will evaporate, leaving the acid behind; this will soon blacken the paper.

2. Pour 1 oz. of strong H_2SO_4 into 2 oz. H_2O, and allow it to cool. Immerse in it a few sheets of dry unsized paper, remove and wash thoroughly. When dry they are very tough and stiff and resemble parchment. Wet a piece of this parchment paper and stretch it over the top of a beaker and tie it tightly. When dry it is perfectly smooth and firm. With the wet finger trace a letter upon a sheet of paper before putting it into the acid; on taking it out the wetted portion drops out (p. 189).

3. Dissolve some sugar in a very little water, so as to form a thick syrup. Put it in a tall beaker and pour on strong H_2SO_4 until it begins to swell up and blacken. The H_2SO_4 removes the H_2O from the sugar, leaving only C behind. (Note, p. 190.)

4. Mix 4 oz. H_2SO_4 and 1 oz. pounded ice or snow. Stir it with a test-tube containing a little ether; the ether soon boils.

5. Mix 1 oz. H_2SO_4 with 4 oz. snow or pounded ice, and stir it with a test-tube containing cold water; the water soon freezes. The vessel in which the experiment is performed usually freezes fast to the table so that it is well to set it on a plate or small board.

* In mixing H_2SO_4 and H_2O always pour the acid *into* the water.

6. Place in the evolution-flask half an ounce of FeS. Cover this with water, and then pour through the funnel H_2SO_4, until the gas comes off freely. It may be passed into a glass of cold water. This solution must be bottled and closely corked. The gas may be tested directly, as mentioned in the text.

7. Add some of the solution to a dilute one of antimony tartrate: a beautiful orange-colored precipitate of antimony sulphide will be separated. With a dilute solution of tin chloride, a yellow, tin sulphide will be formed; and with a solution of copper sulphate, also largely diluted, a brownish-black, copper sulphide will be obtained. (See Chapter on Qualitative Analysis.)

118.—1. Place a few drops of the disulphide in each of four test-tubes. To one add a little powdered sulphur, to a second a few minute scales of iodine, to a third a fragment of phosphorus, and to a fourth a few drops of water. Notice the beautiful color produced by the iodine; the solution of the sulphur and the phosphorus; the insolubility of the liquid in water; and also its refractive power.

120.—1. Cover a stick of phosphorus with dry, finely-powdered charcoal. It will soon ignite.

2. Put in a vial half an ounce of sulphuric ether and a half-dozen pieces of phosphorus not larger than grains of wheat. Thoroughly shake and then set away. Repeat the shaking often. When the phosphorus is dissolved, pour a little of the solution on the hands, and when briskly rubbed together in a dark place they will glow with a ghostly light.

3. Pour some of the solution on a lump of loaf-sugar. Drop this in hot water, when the ether will catch fire.

4. Place a few grains of $KClO_3$ on a brick and pour over them some of the solution. The ether soon evaporates and the mixture explodes.

5. Dissolve 1 or 2 decigrams of phosphorus in 2 c. c. of carbon disulphide in a test-tube; pour a little of the solution upon a piece of filtering-paper, and allow it to dry. The phosphorus will be left in a finely divided form, and will set fire to the paper in a few minutes. The ether solution may be used instead.

6. Place a bit of phosphorus in a solution of silver nitrate. In the course of a day or two it will be covered with brilliant crystals of reduced silver. Repeat with $CuSO_4$ solution.

122.—1. Dissolve 4 grams of caustic potash in 16 grams of water; place it in a small flask of about 50 c. c. capacity, and add 2 or 3 decigrams of phosphorus; immerse the delivery tube just below the surface of water in a small capsule, and heat the mixture gently. Bubbles of gas will form in the retort, and will break with a flash and a slight explosion upon the surface of the potash solution. By degrees the air of the retort will be deprived of all its O, and then the bubbles of gas, as they escape into the air, will take fire, producing a white wreath of phosphoric anhydride, which forms a series of ringlets, revolving in vertical planes around the axis of the wreath itself as it ascends.*

124.—1. Boil 1 gram of arsenious anhydride with three of potassium carbonate in 100 c. c. of water till it is dissolved, and add it to a solution of 3 grams of cop-

* There is danger of breaking the flask by the bursting of the bubbles of gas within it before the air has all passed out. A tea-spoonful of ether placed in the retort before the heat is applied will at once be vaporized, and will carry out the air. Great caution, however, is then necessary, as the ether-vapor is very inflammable. A better way is to fill the flask and its delivery tube entirely with water. Still another method consists in fitting the flask with a second tube reaching to the bottom of the flask, and passing H gas through it both before and after the experiment.

per sulphate in 100 c. c. of water: a beautiful green precipitate of *Scheele's Green* ($CuHAsO_3$) will be obtained.

2. Add a few drops of a solution of arsenious anhydride to 200 or 300 c. c. of water, and then 3 or 4 c. c. of HCl; place in the liquid two or three slips of bright copper foil, and boil the whole for a few minutes: the copper foil will become coated with a steel-gray film. Part of the Cu becomes dissolved, and displaces the arsenic, which is thrown down on the undissolved portion. Pour off the water, dry the Cu on blotting paper, and heat the foil in a tube, sealed at one end. The arsenic will sublime, condensing in minute octahedra on the cold sides of the tube. This is *Reinsch's test* for arsenic. (See note on p. 299.)

3. Place 1 mg. of any As compound upon a fiber of asbestos and bring it into the upper part of a Bunsen gas-flame, which has only a small luminous point to it. Hold a dish of cold water an inch or two above the flame. The As burns to As_2O_3 and is deposited on the porcelain, but being white is invisible. Touch the spot with a neutral solution of $AgNO_3$, then hold the stopper of the ammonia bottle near it, and blow some of the H_3N vapor upon it. A yellow stain appears. Touch a drop of aqua-ammonia to it and it disappears again.

4. Compounds of As when heated on charcoal give off a garlic odor.

127.—1. Before throwing K on water, carefully pare off all the crust that forms on it and remove the adhering naphtha. Pour a gill of water into a deep glass jar, throw in the K, and cover the jar to prevent accidents from flying pieces of the burning metal. View the flame through a piece of dark-blue glass.

2. Burn some dry brushwood; collect the ash, wash it with five or six times its bulk of water, and filter. Test the solution with a piece of reddened litmus-paper, which will become blue. Evaporate the solution to dryness in a small porcelain dish. If the dry mass be left exposed to the air for a few hours it will become moist. The potassium carbonate, of which it chiefly consists, attracts moisture rapidly and deliquesces. To a portion of the salt add a few drops of HCl: brisk effervescence occurs.

3. Place 30 grams of pearlash in a half-litre bottle, and dissolve it in 250 c. c. of water. Shake 20 grams of quicklime with five or six times its bulk of water, and add the pasty mixture (about 120 c. c. in bulk) to the boiling solution of pearlash. Agitate the mixture, and let it stand till it is clear. Pour off a portion of the liquid: it is a solution of caustic potash. Add to it some HCl: no effervescence will occur. Agitate a tablespoonful of olive oil in a small vial with 3 or 4 c. c. of the caustic solution diluted with ten times its bulk of water: a milky-looking liquid will be formed, which is the first stage in the making of soap.

4. Pulverize finely nitrate of potash and chloride of ammonium, five parts of each, and mix with sixteen parts of water. The temperature of the mixture will be reduced so low that if a test-tube with a little water in it be used to stir it, the water in the tube will be converted into ice.

130.—1. Pass Cl gas into a solution of KHO; you will have a solution of potassium hypochlorite, which possesses strong bleaching properties. Continue to pass in Cl until it is no longer alkaline, when crystals of $KClO_3$ will separate.

2. If 4 measures of the cold saturated solution of potassium bichromate be mixed with 5 of oil of vitriol, and the liquid be allowed to cool, chromic anhydride crystallizes in crimson needles, which may be drained and dried upon a brick.

131.—1. The salts of K and Na may also be distinguished in the following manner: To a pretty strong solution of the salt in question add a solution of tartaric acid, and stir the mixture with a glass rod. If K be present, white, gritty crystals of cream of tartar ($KH_5C_4O_6$) will be deposited, but no such precipitate will occur with salts of Na.

2. Make a saturated solution of NaCl and allow it to crystallize slowly; hopper-shaped crystals result. Better crystals are formed if the water contains a little sugar.

3. Make a second solution and add $MgCl_2$ to it; the crystals will be cubical.

4. Take a small saucepan, and, having made a little pool of water upon a wooden stool, set the saucepan upon it; then throw in a handful of snow or powdered ice, and a handful of common salt; now stir with a stick, and the cold will freeze the saucepan to the stool, even before a large fire.

133. Fill a tall cylinder with a clear saturated solution of NaCl, and pass into it at the same time strong currents of H_3N and CO_2 gases. A precipitate of $NaHCO_3$ falls. This is known as the Solvay Ammonia soda process. The solution contains NH_4Cl, from which H_3N may be recovered by treating it with CaO.

137.—1. Place a few lumps of *black* marble in the open fire, or in an open crucible with a hole at the bottom, and heat it strongly for an hour or two. When it is completely converted into quicklime, the lumps, when broken across, will be quite *white*.

138.—1. Into a tumbler of pure water pour a teaspoonful of lime-water and pass CO_2 into it. At first it becomes milky owing to the formation of $CaCO_3$, but the precipitate afterwards dissolves in the excess of CO_2. Pour some of this clear solution into a test-tube and boil; the CO_2 is expelled and the water becomes milky. Such water is said to be "temporarily hard." Fill a second test-tube with the clear solution and add a little lime-water; a precipitate is also produced. Hard water may be softened either by boiling or, paradoxical as it seems, by adding lime-water. Water which contains $CaSO_4$ in solution is said to be "permanently hard," because it cannot be softened in this way.

139.—1. Select a medal suitable for the purpose; paste a shallow rim of paper round it, so as to make it like the lid of a pill-box, and anoint the surface of the medal very lightly with oil. Mix a little of the dry plaster with water till it becomes of the consistence of thin cream; apply it carefully with a hair-pencil to every part of the surface, so as to exclude air-bubbles; then pour a thicker mixture into the mould. Allow it to remain for an hour. The cast may then be removed: it will be a reversed copy of the medal.

141.—1. Place a little of some magnesium salt on a platinum wire moistened with a solution of cobalt nitrate. A pink residue will be obtained on heating the wire in the outer part of a Bunsen gas-flame.

2. Add to a solution of any magnesium salt, such as the sulphate, a solution of potash: a white precipitate of hydrated magnesia is formed. Excess of alkali will not re-dissolve it. Lime-water produces a similar precipitate.

151.—1. Allow a drop of HNO_3 to fall upon a slip of polished steel: a dark gray spot is produced, owing to the solution of the metal in the acid, while the C is left. If the acid be dropped upon a slip of iron a green stain is formed.

2. Take a strip of steel (watch-spring, corset steel, or piece of hoop-skirt), and render it magnetic by rubbing with a magnet (Physics, p. 214). Dip it into iron-filings and hold them in an alcohol flame. They take fire and burn.

3. Pulverize a salt of iron and heat it with Na_2CO_3 on charcoal or on a match stick (p. 257). A black powder (Fe_3O_4) is obtained, which is attracted by the magnet.

155.—1. Heat a bar of iron white-hot, and bring it in contact with a roll of S over a pail of cold water. The S and Fe immediately unite, and form drops of a reddish-brown color, which fall into the water. This is ferrous sulphide, FeS, useful in making H_2S for laboratory purposes.

2. Potassium permanganate ($K_2Mn_2O_8$) may be obtained by mixing 40 grams of finely-powdered manganese dioxide with 35 grams of potassium chlorate, and adding a solution of 50 grams of caustic potash to the mixture, evaporating to dryness, and heating the powdered residue to dull redness in a clay crucible. When cold, the mass is treated with water, and decanted from the insoluble residue; a splendid purple liquid is obtained, which on evaporation yields needles of the permanganate.

3. Pulverize a few crystals of permanganate and dry them in a capsule held for a moment over the flame. Pour on them a few drops of the strongest H_2SO_4. Bring the dish in front of an H generator or direct against it a fine jet of illuminating gas. The gas takes fire from the ozone liberated.

4. Make a strong solution of the permanganate and heat to boiling in a test-tube. Pour in a few drops of glycerine; the latter is oxidized so violently that a flash may be seen, and part of the liquid is ejected from the test-tube.

5. *The permanganate test for impure water.* Take 4 parts of potassium permanganate and 4 parts of potassium hydrate, and dissolve in 160 parts of freshly-distilled water. One minim of this solution placed in a test-tube of distilled water, remains of a beautiful pink hue for several days, but if the minutest trace of egg albumen be added to the same quantity of water, it will be infallibly detected. So, if on the addition of a minim of this solution to a wine-glass of water from any well used for drinking purposes, the water, after a few hours, gives a brownish precipitate with loss of color, such water contains an abnormal proportion of organic matter, and is injurious to health.

157. Make a transparent solution of zinc sulphate (white vitriol) so abundantly formed in preparing H gas. Take a glass half-full of this solution and another half-full of strong ammonia. Pour them together, and if the proportion be properly calculated, the two liquids will form a solid so dry, apparently, that on inverting the glass containing it, not a drop will fall out. Cold will be produced by this change from a liquid to a solid state. (See Physics, p. 187.)

158.—1. Fill a test-tube nearly full of H_2O. Pour in it a few drops of the solution of copper sulphate. Add H_3N, and a blue precipitate will be formed. Notice the change from green to blue. The copper sulphate may be readily prepared for this experiment by covering a copper cent with dilute oil of vitriol. This experiment may be made to show the divisibility of matter by weighing the cent, finding what proportion of the whole solution you use, and then experiment to see what quantity of water can be taken and yet have the blue color perceptible in the ammonia test.

2. Besides the ammonia test for copper, the metal may be detected, (1) by the red metallic deposit formed on a polished plate of iron if dipped into a solution of the salt, (2) by the black insoluble sulphide produced by H_2S, and (3) by the blue hydrate turning black on heating.

160.—1. If a water contain lead, even in minute quantity, its presence is easily ascertained by taking two similar jars of 25 c. m. high, of colorless glass, filling both of them with the water, and adding to one of the jars 3 or 4 c. c. of a solution of sulphuretted hydrogen. A quantity of lead less than one part in two millions is easily perceived by the brown tinge occasioned, on looking down upon a sheet of white paper; the jar to which the test has not been added serving as a standard of comparison.

162.—1. Place a little gold leaf in two test-tubes; to one add HNO_3, to the other HCl. Even when heated, the gold leaf will remain unaffected in each. Pour the contents of one tube into the other: the Au will disappear with effervescence. Evaporate this solution in a small porcelain dish till the acid is nearly all driven off: gold chloride will be left.

2. Dilute the solution with 3 or 4 c. c. of H_2O. To a portion of this liquid add a solution of ferrous sulphate: a brown precipitate of finely divided reduced Au is obtained, and iron chloride is formed.

167.—1. Dissolve a ten-cent-piece in HNO_3. The solution has a bluish color, owing to the presence of the Cu. Dilute with 200 c. c. of water; then add a solution of NaCl so long as it forms a precipitate; white flakes of silver chloride are formed. Stir the mixture briskly with a glass rod; the precipitate will collect into clots. Filter the solution. The presence of Cu may be found in the clear liquor by adding to a portion of the liquid H_3N in excess: a blue solution is formed. Place the blade of a knife in another portion of the filtrate; it will become coated with metallic Cu.

2. Take the precipitated silver chloride, and after having washed it well on a filter, place it in a wine-glass with a little water; add two or three drops of H_2SO_4, and then place a slip of Zn in contact with the chloride, and leave it for twenty-four hours. The chloride will be reduced to metallic Ag, which will have a gray porous aspect, while zinc chloride will be found in solution. Lift out the piece of Zn carefully; wash the Ag first with water containing a little H_2SO_4, then with pure H_2O. Dry the residue. Place a small quantity of it upon an anvil, and strike it a blow with a hammer; a bright metallic surface will be produced. Place a little of the gray powder upon charcoal, and heat in the flame of a blow-pipe: it will melt into a brilliant malleable bead. Dissolve another portion in HNO_3; red fumes will escape, and silver nitrate be obtained in solution.

3. Fill a vial half-full of a solution of silver nitrate and add a few globules of Hg. The Ag will be precipitated in a few days, forming the "silver tree."

4. Place a crystal of $AgNO_3$ on a piece of charcoal and heat in the reducing blow-pipe flame. The coal becomes covered with a film of metallic Ag.

5. Float a sheet of sized paper on an NaCl solution. When dry, float it for 3 minutes on a solution of $AgNO_3$ and dry in the dark. Press a fern leaf on a piece of glass, lay a sheet of this paper on it and then a thin board as large as the glass. Clamp them together with clothes-pins, and expose to direct sunlight. When sufficiently black place in a solution of "hypo," and wash for 24 hours.

6. In performing the experiment given at the bottom of p. 167, do not allow the solution to remain in the test-tube, but throw it away at once, as the black precipitate deposited, ***when dried***, is powerfully explosive and is known as fulminating silver.

191.—1. Put a little $AgNO_3$ solution in a test-tube and add aqua-ammonia slowly until the brown precipitate has again dissolved. Pour some of this into a second test-tube and add a solution of grape-sugar. On boiling, the silver will be reduced in the form of a brilliant mirror on the side of the tube.

2. Take a solution of $CuSO_4$ and add enough tartaric acid to prevent its being precipitated by KHO. Then add enough KHO solution to make it strongly alkaline. Pour some of this solution into a dilute solution of grape-sugar and boil.

The intensely blue color disappears and a red precipitate of Cu_2O is produced. This is called *Fehling's test*, and is employed to detect sugar in diabetic urine.

192.—1. To one gill of water add fifteen or twenty grains of strong H_2SO_4. Place in a large flask and heat. While boiling, drop in slowly two drams of starch, finely powdered. Boil for several hours, adding water as may be necessary. Finally, drop in slowly fine chalk until the liquid is neutral; then cool, filter off the calcium sulphate, and evaporate the liquid to a syrup.

193.—1. Dissolve 1 oz. grape-sugar in 10 oz. of water and add a small quantity of yeast. Place the mixture in a flask connected by means of a glass tube with a second flask (Fig. 45) containing lime-water. Let it stand 24 hours in a warm place. The lime-water will become milky from the formation of $CaCO_3$. The contents of the flask will have an alcoholic odor, and by distilling, a dilute alcohol could be obtained.

197.—2. Mix together *slowly* 2 oz. of alcohol and 2 oz. of H_2SO_4, and heat to 150° C. in a retort connected with a cooler. *Ethylic ether* distills over.

2. After washing out the retort used in the last operation, introduce into it 3 oz. potassium acetate, 3 oz. alcohol, and 2 oz. H_2SO_4. Mix the alcohol and acid, and pour upon the pulverized salt; heat gently. *Ethyl acetic ether* is formed.

3. Heat a mixture of 2 oz. potassium acetate, 1 oz. H_2SO_4, and 1 oz. amylic alcohol. An artificial *oil of pears* is produced.

199.—1. Pour a little alcohol into a small beaker and suspend over it a strip of Pt foil which has been heated until it glows. It will continue to glow owing to a slow oxidation of the alcohol vapors. The peculiar penetrating odor observed is due to *aldehyde.*

200.—1. Marsh gas, CH_4 (p. 81), is made by heating a mixture of 2 parts sodium acetate, 2 parts KHO, and 3 parts powdered quicklime in a tube of hard glass or a metallic retort, a high temperature being required.

2. Heat a mixture of NaCl, methylic alcohol, and H_2SO_4. A colorless gas, CH_3Cl, is obtained, which burns with a beautiful *green* flame.

206.—1. Mix together in a test-tube equal quantities of H_2SO_4 and HNO_3, and when cold, allow some benzole * to flow into it, drop by drop. On pouring the mixture into water it will be found that the benzole no longer floats on water as before, and has acquired a very agreeable odor. An atom of H has been replaced by NO_2, forming the poisonous *nitro-benzole*, $C_6H_5NO_2$.

214.—1. Dissolve 1 grain of quinine sulphate, in water acidulated with H_2SO_4. It exhibits in a dark room a beautiful fluorescence. Add chlorine-water and then aqua-ammonia; a bright green color is produced, characteristic of this alkaloid.

216.—1. Fill three tall cylinders nearly full of water; to one add enough cochineal solution to color it faintly pink, to a second a little litmus. To each add some aqua-ammonia and then a solution of alum. In 24 hours a pink precipitate is seen in the first, a blue in the second, and a white in the third, while the liquid in each has become colorless. These precipitates are called *lakes.* Dirty water becomes clear when treated with alum and ammonia.

219.—1. Fill a test-tube one-sixth full of sweet oil, add a little ammonia, and nearly fill with water. The constituents remain separate. Shake thoroughly,

* Benzine from petroleum, does not possess this power of forming nitro compounds.

and they will combine, forming a thin, soapy liquid. Add an acid, and they will separate at once.

228.—The following experiment shows that albumen in the presence of starch is not coagulated, even at a boiling temperature. It makes a fine class illustration of catalysis. Mix 50 grains of pure starch with 1 fluid-ounce of water. Dilute the albumen of one egg to make 3 fluid-ounces and strain through muslin. Mix the two solutions and boil. There will be no coagulation or precipitate. Filter. Add a drop of nitric acid to the clear liquid, and instantly a dense white coagulation will be formed.

229.—1. Fill a test-tube one-third full of fresh milk and add an equal volume of water, then a little acetic acid (or rennet), and allow it to stand a short time; filter out the precipitate and wash with water. The precipitate consists of casein and fats. The filtrate contains sugar, as may be shown by its reducing action on an alkaline solution of tartrate of copper. Remove the precipitate from the filter and shake it up in a test-tube with ether. This dissolves out the fat. Filter again; let the filtrate evaporate and the fat will be left in a pure state.

2. Evaporate a larger portion of milk to dryness, and heat until the residue is quite white. Dissolve in water, filter and test for $NaCl$ with $AgNO_3$.

230.—1. Mix intimately 10 grains of gelatine with 50 grains of soda-lime, and heat strongly; NH_3 gas will be evolved and can be detected by its odor, its action on reddened litmus-paper, and by fuming with HCl.

2. Make a dilute solution of gelatine and add a solution of tannic acid; a leathery precipitate is formed (p. 211).

3. Mix together equal volumes of a solution of gelatine and a solution of $K_2Cr_2O_7$, and add a little tincture of logwood. Pour it into a flat dish and float upon it a sheet of unruled writing-paper and dry in the dark. Expose under a negative (p. 168) or fern leaves, to direct sunlight for an hour, then wash thoroughly with *warm* water. Bichromated gelatine is rendered insoluble by exposure to sunlight. This principle is employed in all photo-engraving processes.

233.—A number of instructive experiments may be shown to prove the power of the sunbeam.

1. Mix together equal volumes of dry Cl and CO, and expose to direct sunlight. They combine quietly without explosion to form a colorless gas of pungent, suffocating odor, called *Phosgene* gas. The volume of gas formed is only half that of the gases employed.

2. Mix together equal volumes of H and Cl, in the dark, fill a small glass bulb, and throw it up into a bright beam of the sun, when it explodes *with violence.* Do not at any time hold the bulb in the hand or within several feet of the eyes.

3. Plants decompose CO_2 in the sunlight. (See p. 99.)

4. The ordinary photographic process, as given on p. 168, is a good illustration of the power of the sun's rays; also that given with gelatine on this page.

5. Dissolve 1 gram of ammonia-citrate of iron in 20 c. c. of water; add to it 20 c. c. of a solution of K_4FeCy_6 made in the same manner. Keep the mixed solution in the dark. Float a sheet of white paper on the solution and allow it to dry. Cover a plate of glass with black varnish, and before it dries write upon it with any sharp-pointed instrument. When perfectly dry, place it over a sheet of the prepared paper and expose two or three hours to bright sunlight. Remove and wash well in cold water. You will have the writing in blue on a white ground.

QUALITATIVE ANALYSIS,

FOR BEGINNERS.

[The following pages on analysis were prepared by Prof. E. J. HALLOCK, Ph. D., of Boston.]

IN order to be able to analyze almost every inorganic substance met with in the arts, or sold in the shops, it is only necessary for the student to familiarize himself with the reactions of about twenty-six metals and a dozen acids. To be able to apply these tests with certainly, in all cases, and to know the easiest and best methods of dissolving the substance, constitute a qualitative chemist.

For reasons which will appear farther on, metals are divided into five groups.

THE FIRST GROUP embraces lead, silver, and the suboxide salts of mercury. They are classed together because they are the only metals whose chlorides are insoluble in acids. The student should take a solution of lead nitrate $Pb(NO_3)_2$, formed by dissolving litharge in nitric acid, or some lead acetate solution (see page 161), and try the following tests, making a note of his results. With HCl a white precipitate of $PbCl_2$ is formed. This precipitate is filtered out, washed, and dissolved in boiling water. To another portion of the solution add H_2SO_4, a white precipitate of $PbSO_4$, insoluble in H_2O. To a third portion add potassium chromate, K_2CrO_4, a yellow precipitate is formed; to another portion add KI; a yellow precipitate is again produced, but it dissolves in boiling water, and forms beautiful crystals on cooling.

Repeat each of these tests with silver nitrate, $AgNO_3$ (*Experiment 167*).* The precipitate with HCl is insoluble in boiling H_2O, but dissolves in NH_4HO. With KI it forms a yellowish-white precipitate. Repeat with mercurous nitrate, $HgNO_3$, made by the action of very

* These numbers refer to the chap. of Directions about Experiments in the Appendix, p. 245.

dilute HNO_3 on an excess of Hg, in the *cold.* We have with HCl a precipitate of calomel (HgCl), (page 172), which is insoluble in H_2O, and blackens on adding NH_4HO, but does not dissolve. KI forms a greenish-yellow precipitate.

SEPARATING METALS OF GROUP I.—Mix the solutions of the three metals, and add HCl. Filter, and boil the precipitate in water; filter hot, and to the *filtrate* add K_2CrO_4. The yellow precipitate proves that lead is present. Boil the residue in ammonia and filter; to the filtrate add HNO_3. A white precipitate proves silver present. The black insoluble residue is a compound of mercury (Hg_2H_2NCl).

THE SECOND GROUP embraces the protoxide salts of mercury, together with Pb, Bi, Cu, Cd, As, Sb, Sn, Au, and Pt. They are precipitated from acid solutions, by H_2S gas being passed through the solution, as sulphides. (See *Experiment 117.*) Of these HgS, PbS, Bi_2S_3, CuS, and CdS are insoluble in ammonium sulphide $(NH_4)_2S$, and constitute the first division of this group. The sulphides of the remaining five metals are soluble in $(NH_4)_2S$, and form the second division. $(NH_4)_2S$ is prepared, according to Fresenius, by saturating a given volume of ammonia solution (specific gravity 0.96) with H_2S gas, and adding to it an equal volume of the ammonia. The solution, which is at first colorless, soon becomes yellow by keeping, or may be at once converted into the yellow sulphide by the addition of sulphur. It should yield no precipitate with magnesium sulphate (Epsom salt). This reagent is decomposed by acid, sulphur being precipitated.

Pass H_2S gas into a solution of corrosive sublimate ($HgCl_2$); a precipitate is formed which is at first white, then yellow, red, and finally black. It is insoluble in $(NH_4)_2S$, and in HNO_3. When dissolved in aqua regia it gives a grey precipitate with $SnCl_2$. Repeat the first experiment with some lead solution; a black precipitate is formed, soluble in boiling HNO_3. In this solution a white precipitate of $PbSO_4$ is formed on adding H_2SO_4. Add a few drops of KI solution to the original solutions of $HgCl_2$, and $Pb(NO_3)_2$; in the former case the precipitate (HgI_2) is red, in the latter (PbI_2) it is yellow. These tests are characteristic of the metals when alone. (*Experiment 107.*) Pass H_2S into $Bi(NO_3)_3$ solution;* a black precipitate; dissolve in HNO_3; add a drop of H_2SO_4 to prove it is not lead; then cautiously add ammonia, which produces a white precipitate of

* When Bi solutions are diluted with water, basic salts are precipitated unless there be too much free acid present. This reaction is most sensitive with $BiCl_3$, so that HCl may be used to dissolve the $Bi(NO_3)_3$ for the H_2O test.

$Bi(HO)_3$. Repeat all the above experiments with $CdSO_4$ solution; the precipitate with H_2S is a beautiful yellow, soluble in HNO_3, but insoluble in KCy. Pass H_2S in $CuSO_4$ solution, and a brownish-black precipitate will be formed, soluble in HNO_3 and in KCy. Salts of copper have a bluish color, which becomes more intense on adding NH_4HO. (*Experiment 158.*) With potassium ferrocyanide (K_4FeCy_6) they give a reddish-brown precipitate insoluble in HCl.

SEPARATING METALS OF SECOND GROUP, FIRST DIVISION.—After the student has made all the above reactions he may mix the solutions of the five metals and proceed to separate them. Some of the lead is precipitated by HCl, and is filtered out before H_2S is passed through the solution. The precipitate with H_2S is boiled in HNO_3, and HgS remains as a residue. When the solution is very acid, part of the H_2S is decomposed and S precipitated, which must not be mistaken for HgS. To the filtrate add a little H_2SO_4 to precipitate any lead present; filter and add NH_4HO, when $Bi(HO)_3$ is precipitated. The precipitate, dissolved in aqua regia and concentrated by evaporation, should give a white precipitate if poured into water. The addition of NH_4Cl aids this reaction. The *blue* filtrate from the Bi precipitate is boiled with KCy, care being taken not to inhale the fumes, and H_2S added; CdS forms a yellow precipitate. The presence of Cu in the filtrate is proved by the formation of a reddish-brown precipitate with HNO_3 and K_4FeCy_6.

The most interesting metal of group second, second division, is As. H_2S in an HCl solution of As_2O_3 gives a yellow precipitate As_2S_3; it is soluble in $(NH_4)_2S$, and in $(NH_4)_2CO_3$. The neutral salts of arsenious acid yield with $AgNO_3$, yellow precipitates, Ag_3AsO_3 soluble in HNO_3. A small piece of bright green wall-paper usually contains enough of this metal to give several characteristic tests. Apply a single drop of nitric acid to the paper; a moment after neutralize with ammonia and observe the color, a deep blue always indicating copper. When the white fumes have nearly disappeared, apply to the same spot a drop of $AgNO_3$; a yellow ring indicates As. The most delicate test for As as well as Sb is Marsh's test (see page 124). The mirror formed by As on porcelain is soluble in bleaching powders (see page 106), sodium hypochlorite, also called Labarraque's solution (see note, page 104), etc.; that of Sb is insoluble in these (see note, page 173). If AsH_3 is passed into $AgNO_3$, metallic Ag is precipitated and enough HNO_3 is set free to keep the yellow Ag_3AsO_3 in solution until it is boiled with sodium acetate, when the precipitate reappears. (*Experiment 124.*)

Antimony closely resembles As in its reactions. Pass H_2S into a solution of tartar emetic (*Experiment 117*, 7), and an orange-colored precipitate, Sb_2S_3, will be formed, soluble in $(NH_4)_2S$, and on adding dilute HCl to this solution it is reprecipitated. This precipitate is soluble in strong HCl, while the corresponding arsenic precipitate is not, Put some of the Sb solution in a new Marsh's apparatus. The mirror is insoluble in bleaching powders.

Tin dissolves in HCl, but is oxidized to a white powder by HNO_3 without dissolving. There are two series of tin salts; $SnCl_2$ gives a black precipitate with H_2S; $SnCl_4$ a yellow precipitate with H_2S, both soluble in yellow $(NH_4)_2S$. $AuCl_3$ is a test for Sn. $SnCl_2$ forms with an excess of $HgCl_2$, a white precipitate of HgCl; but when $SnCl_2$ is in excess a gray precipitate of Hg is formed. On adding dilute HCl to the $(NH_4)_2S$ solution of the sulphide, it is reprecipitate yellow, even though it may have been black before it was dissolved. It dissolves in strong HCl.

Gold and platinum are distinguished from all other metals by their insolubility in HCl or HNO_3, but are converted into soluble chlorides by aqua regia. The characteristic test for Au salts is $SnCl_2$ mixed with $SnCl_4$, the purple of Cassius being formed. $PtCl_4$ will give a yellow precipitate with NH_4Cl and alcohol, $(NH_4)_2PtCl_6$.

SEPARATING METALS OF SECOND GROUP, SECOND DIVISION.—Into an acid solution of Sn, Sb, and As, pass H_2S gas; filter and dissolve precipitate in $(NH_4)_2S$ to remove any members of first division present. Reprecipitate the sulphides with dilute HCl, filter and wash. Then dissolve in strong HCl; if a residue remains it is probably As_2S_3. Test this as described in *Experiment 124*, 3 (page 267), or in Marsh's apparatus. The soluble portion contains Sb and Sn. Put this in another Marsh's apparatus; the mirror will be insoluble in hypochlorites. When the zinc has all dissolved, take the residue and dissolve it in HCl and tested for Sn with $HgCl_2$.

As Au and Pt are seldom to be sought for, this part of the separation may be omitted. The mixed sulphides of Au, Pt, As, Sb, and Sn may all be placed in Marsh's apparatus, as directed in the table on page 288, when the As and Sb combine with H, and are separated by passing the AsH_3 and SbH_3 into $AgNO_3$. The metallic Sn, Au, and Pt remain in the H generator; the Sn is then dissolved out with HCl, and tested with $HgCl_2$; the Au and Pt are dissolved in aqua regia and tested in separate portions of the solution as above described.

GROUP THIRD embraces Co, Ni, Fe, Cr, Mn, U, Al, and Zn. They are precipitated by $(NH_4)_2S$ from neutral or alkaline solutions.

The characteristic test for Co, is the blue color imparted to a borax bead. Ni alone gives, in the outer blow-pipe flame, a reddish-brown bead, in the inner gray. Both give with $(NH_4)_2S$ black precipitates insoluble in dilute HCl. If KNO_2 and acetic acid are added to a solution of Co and Ni, the former is slowly precipitated and not the latter, To a solution of $FeSO_4$, add a drop of potassium ferricyanide ($K_6Fe_2Cy_{12}$); a blue precipitate is formed. To another portion add $(NH_4)_2S$; a black precipitate is formed, soluble in dilute HCl, from which solution it is re-precipitated by NaHO, as $Fe(HO)_2$. Repeat the latter test with ferric chloride (Fe_2Cl_6) and the same result is obtained. Fe_2Cl_6 gives with K_4FeCy_6 a precipitate of Prussian blue. Into a glass of water place one drop of Fe_2Cl_6, and add potassium sulphocyanide (KCyS); the liquid acquires a beautiful red color. Iron salts also give characteristic colors to the borax beads. To a solution of $MnSO_4$, add $(NH_4)_2S$; a flesh-colored precipitate is formed, soluble in HCl, re-precipitated on boiling in NaHO, as $Mn(HO)_2$. The borax bead with Mn acquires an amethyst-red color (see note, page 109) in the outer blow-pipe flame. Fused with Na_2CO_3 and KNO_3 a green mass is formed. $(NH_4)_2S$ gives with the salts of Cr, a greenish precipitate of $Cr_2(HO)_6$, soluble in HCl, and re-precipitated by NaHO. Fused with Na_2CO_3 and KNO_3 it forms the yellow potassium chromate K_2CrO_4. When this is dissolved in water and acidified with acetic acid, it yields a yellow precipitate with $Pb(NO_3)_2$. To a solution of alum add $(NH_4)_2S$; it forms a white precipitate soluble in dilute HCl. To the second portion add NH_4HO, a white precipitate; and to a third add NaHO and boil; the white precipitate at first formed dissolves again. To a solution of $ZnSO_4$, obtained in *Experiment 50*, page 252, add NH_4HO slowly. The white precipitate at first formed dissolves when more NH_4HO is added. In this solution H_2S causes a white precipitate of ZnS.

SEPARATION OF METALS OF GROUP THIRD.—Some NH_4Cl and NH_4HO is first added (see page 274), then $(NH_4)_2S$. The precipitate is digested in dilute HCl; Ni and Co are sought in the residue. The filtrate is boiled with NaHO for half an hour in a porcelain capsule, and filtered. The filtrate is divided in two portions, to one HCl is added until acid, then NH_4HO, which precipitates Al; to the other H_2S to precipitate the Zn. The residue, containing Fe, Mn, and Cr, is fused with pure KNO_3 and Na_2CO_3; if the mass is green, Mn is indicated; if yellow, Cr. One-half of the mass is dissolved in water; the insoluble residue is tested for iron; the filtrate is tested for Cr by first neutralizing with acetic acid, and then adding $Pb(NO_3)_2$. The test for Mn is to place some of the fused mass in HNO_3, with red lead; if left at rest, a beautiful rose pink is formed from the conversion of K_2MnO_4 into $K_2Mn_2O_8$. (*Experiment 155*, 2.) U is rarely sought.

GROUP FOURTH embraces the metals of the alkaline earths, Ba, Sr, and Ca, whose carbonates, precipitated by $(NH_4)_2CO_3$ are insoluble in H_2O, but soluble in HCl. $BaCl_2$ forms with H_2SO_4 a precipitate insoluble in acids; it is also precipitated yellow by K_2CrO_4. Ba compounds impart a green color to the flame of an alcohol lamp or Bunsen burner. $CaCl_2$ with ammonium oxalate, yields a white precipitate insoluble in water. H_2SO_4 produces a white precipitate of $CaSO_4$, slightly soluble in water and acids. Ca salts color the flame yellowish red. $SrCl_2$ gives a white precipitate with a clear solution of $CaSO_4$; if the solution is dilute, half an hour is required for the precipitation. Sr colors the flame carmine-red.

SEPARATING METALS OF GROUP FOURTH.—Some NH_4Cl is first added, if not already present in the solution, then $(NH_4)_2CO_3$. The precipitate is dissolved in acetic acid, and divided in two portions. To one is added K_2CrO_4; Ba is precipitated yellow, and the filtrate is tested for Sr by means of $CaSO_4$. H_2SO_4 is added to the other portion, the precipitate filtered out. NH_4HO is added to the filtrate, which is now tested for Ca with ammonium oxalate.

GROUP FIFTH embraces Mg, K, Na, and L. As lithia is very rare, we omit its reactions. $MgSO_4$ yields a white precipitate of $Mg(HO)_2$ on the addition of NH_4HO, unless the solution contains NH_4Cl; hence the necessity of adding NH_4Cl, before testing for Groups III and IV, where NH_4HO would otherwise throw down Mg. The salts of Mg give a white precipitate with hydrodisodic phosphate, Na_2HPO_4 and NH_4HO. KCl yields a yellow precipitate with $PtCl_4$. Potassium acetate is also precipitated in concentrated solutions by tartaric acid. K_2SO_4 gives a white precipitate with $2HF,SiF_4$ and alcohol. K imparts a violet color to flame, which appears red when viewed through blue glass, Na is not precipitated by any of the above re-agents; but may give with $NaSbO_3$ a white precipitate. It imparts an intense yellow color to flame. K, Na, and L, as well as Ca, Ba, and Sr are easily detected by the spectroscope.

Ammonia is liberated from its compounds by mixing with NaHO or $Ca(HO)_2$, and is then recognized by the smell, by bluing red litmus, and by producing white fumes when a rod moistened with HCl gas is held over it, (See pages 49, 135.)

TESTS FOR ACIDS.

The acids do not admit of the strict grouping and successive separation employed for metals, and we will rest content with mentioning the simplest tests for the principal acids, beginning with the haloids :

HCl with $AgNO_3$, white precipitate, soluble in NH_4HO, not in HNO_3.

HI " " yellowish precipitate, insoluble in NH_4HO.

HI " $HgCl_2$, red precipitate, soluble in KI.

HI " starch paste and Cl solution or bleaching powders, blue color. (See page 107.)

CaF_2 with H_2SO_4 liberates HF, which attacks glass. (See p. 106.)

HBr " starch paste and Cl water, an orange-yellow color.

H_2SO_4 with $BaCl_2$, white precipitate, insoluble in HCl.

SiO_2 is insoluble in H_2O, as are most of the silicates except those of K and Na. In analyzing the soluble silicates, they are first evaporated to dryness with excess of HCl, the soluble chlorides dissolved in H_2O or HCl, and the SiO_2 left as a gritty powder.

Boracic Acid is detected by placing it in a capsule containing alcohol and H_2SO_4, and igniting the alcohol. A green tinge to the flame indicates Bo. If a mixture of glycerine and borax is brought into a flame and removed as soon as it takes fire, the green flame is easily recognized. (*Experiment 112.*)

H_3PO_4 with neutral $AgNO_3$, yellow precipitate, soluble in HNO_3 and NH_4HO.

H_3PO_4 with solution of ammonium molybdate in HNO_3, a fine yellow precipitate.

H_3PO_4 with $MgSO_4$ solution containing NH_4Cl and NH_4HO, a white precipitate soluble in acids. (See test for Mg, page 278.)

CO_2 Carbonates effervesce on the addition of acids, CO_2 being set free, which extinguishes a match inserted in the test tube. The ear is often able to detect slight effervescence not otherwise perceptible. (See page 74.)

HNO_3 is not precipitated by any reagent. Into a test-tube containing some nitrate, drop a crystal of $FeSO_4$, then allow a drop of H_2SO_4 to flow down the side of the test-tube which is held inclined. A characteristic dark-brown ring forms immediately. If Cu and strong H_2SO_4 are heated with a nitrate, red fumes are given off. A nitrate heated on charcoal deflagrates.

Chlorates deflagrate more violently than nitrates. H_2SO_4 liberates ClO_2 which is betrayed by its color and odor. If a crystal of $KClO_3$ and a piece of P be placed in a glass of water, and a drop of H_2SO_4 conveyed to it by a pipette or tube, the P takes fire and burns under water (page 130, *Experiment 2*). All experiments with chlorates must be performed with minute quantities, because of the great danger of explosions.

SO_2 is easily recognized by its odor. When sulphites are treated with HCl, the SO_2 is evolved.

If Cl gas be given off on heating a substance in HCl, the presence of a binoxide may be suspected. (See page 103.)

If an odor of H_2S is perceived on treating a substance with HCl, it is evidently a sulphide. Heated with strong HNO_3, sulphides are converted into sulphates.

HCy with $AgNO_3$, white precipitate soluble in KCy ; difficultly soluble in NH_4HO. Care must be taken in handling the poisonous cyanides. On adding HCl to a cyanide, HCy is liberated, and is detected by the odor, which resembles bitter almonds.

H_4FeCy_6 with $AgNO_3$, white precipitate insoluble in NH_4HO, and in HNO_3. With Fe_2Cl_6, Prussian blue [$Fe_4(FeCy_6)_3$] is formed.

Oxalic Acid, $H_2C_2O_4$, yields a white precipitate with $CaCl_2$, which is insoluble in NH_4Cl, but soluble in acetic acid.

When several acids are present they are tested for in separate portions. As_2O_3 may occasionally be mistaken for H_3PO_4. $BaCl_2$ will yield precipitates with carbonic, oxalic, phosphoric, and sulphuric acids, but they all dissolve in HCl except $BaSO_4$.

To test for HCl in the presence of HI, or HBr, or both, the dry powder is mixed with dry $K_2Cr_2O_7$, and pure concentrated H_2SO_4 added, and heated, when CrO_2Cl_2 is given off as a brownish-red gas ; and a glass rod dipped in NH_4HO and held over the tube becomes slightly yellow if Cl is present, from the formation of $(NH_4)_2CrO_4$. If the substance to be tested contains only Br and I, these two may be separated by $CuSO_4$ and H_2SO_3, which precipitates the latter as Cu_2I_2 and permits the use of the starch test for Br. Or the solution is mixed with CS_2 and a very little Cl water, or solution of bleaching powder, is then poured in and test-tube well shaken. The I is liberated and dissolved by the CS_2, imparting to it the well-known purple color, the intensity of which conceals the yellowish-brown color of the Br likewise set free and dissolved. On adding more Cl water, the violet disappears, leaving only the Br to color the ether.

PRELIMINARY TESTS.

A few tests in the dry way will give some clew to the substances present; but in a complete analysis *every* acid and *every* metal must be sought for.

I. Heating in a Tube of Hard Glass closed at one end.—If the substance blackens, organic matter is present. If vapors escape, they are tested for CO_2, SO_2, H_2S, etc. If a sublimate is formed, it may be S, Hg, or a compound of As or Sb.

II. Heating on Charcoal.—A small portion of the substance is placed on charcoal and exposed to the inner blow-pipe flame. (See page 69.) If an infusible white residue remains, moisten it with $Co(NO_3)_2$; a fine blue indicates Al, a reddish tint Mg, a green color Zn. Mix another portion with Na_2CO_3 and heat on charcoal in the reducing flame. If a metallic globule is formed without an incrustation, it indicates Au, Cu, or Ag, as the color is yellow, red, or white. A very fusible and malleable globule surrounded by a yellow incrustation indicates Pb; if the incrustation is white it may be Sn, and on moistening with $Co(NO_3)_2$ the incrustation turns green. Bi and Sb may be reduced to brittle metallic globules. If As is present, an odor resembling garlic is noticed. The charcoal tests will be of little use to the student, except for detecting Ag and Pb, until practice has given him considerable facility in the use of the blow-pipe. (See *Experiment 69*, 6, page 257.)

III. Borax Beads.—Several metals impart characteristic colors to borax glass when fused with it before the blow-pipe. The end of a piece of platinum wire is bent to form an eye as large as this letter O; it is next dipped in borax and held in the flame until fused, then dipped in the powdered substance and fused again. The color varies according as the oxidizing or reducing flame is employed. (See page 92.)

Color.	Oxidizing.	Reducing.
Blue	Co and Cu	Co
Green	Cr and Cu	Fe and Cr
Red	Fe and Ni	Cu (opaque)
Amethyst	Mn	——
Yellow to brown	Fe	——
Colorless	Si, etc.	Mn
Grey	——	Ni

SOLUTION.

The first thing to be done before beginning an analysis is to bring the substance into solution. Distilled water is first employed; if a residue insoluble in water remains, it is treated with acid. In analyzing metals and alloys, nitric acid is the usual solvent; aqua regia being required only for the noble metals. If Sn is present, and Pb and Ag absent, HCl is employed. Mineral substances, if insoluble in any acid, are rendered soluble by *fluxing*, or fusing with pure Na_2CO_3 and K_2CO_3. As a very high heat is required for fluxing, *deflagration* is sometimes preferred. One part of the insoluble powder is intimately mixed with two parts of dry sodium carbonate, two parts pulverized charcoal, and twelve parts nitre. The mixture is placed in the open air and a match applied. A portion of the porous mass produced will be soluble in water, the remainder in acids. The two solutions are to be preserved and tested separately. The metals will be found in the acid solution, while the acids will be found in the aqueous solution. Before beginning the regular course of analysis with these solutions, part of the aqueous solution is evaporated to dryness with excess of HCl to render all the SiO_2 insoluble. In separate portions of the aqueous solution, the various acids are sought as above described (p. 279).

If a portion of the substance is insoluble in HCl after fluxing, it is probably silicic acid, or an undecomposed silicate, and may be rendered soluble by fluxing a second time.

A platinum crucible must never be employed if reducible metals, especially Pb, As or Sb have been found in the preliminary tests.

EXAMPLES FOR PRACTICE.

AFTER the student has made all the tests above given, and succeeded in separating the members of each group from each other, especial care being given to the separation of lead from bismuth, copper from cadmium, arsenic from antimony, and nickel from cobalt, the teacher may give out the following or similar substances for analysis, not following the precise order of the book, so that the student shall not know what substance he is analyzing. Each student should record the results of every analysis in a note-book which he will rule for each analysis as shown under No. 1.

1. ANALYSIS OF $CuSO_4$.—A crystal of this salt as large as a pea is given to a student, who dissolves it in distilled water in a tes-tube and divides the solution in three portions. To one is added a drop of HCl, which should produce no precipitate. H_2S is then added until all the Cu is precipitated. It is then filtered and the precipitate thoroughly washed on the filter. H_2S should produce no precipitate in the filtrate. The precipitate being insoluble in $(NH_4)_2S$, is dissolved in HNO_3, and no residue, except perhaps a little sulphur, remains, so that the absence of Hg is established. H_2SO_4 produces no precipitate in this solution, neither does NH_4HO; hence Pb and Bi are also absent. The ammonia, however, gave the intense blue color characteristic of Cu, and as only one metal is to be sought, the presence of Cu is farther proved by adding HNO_3 and K_4FeCy_6, which causes a reddish-brown precipitate. The second portion of the solution is used in testing for acids. To a small quantity of this some $BaCl_2$ is added, and if the precipitate is insoluble in HCl, the acid present must be H_2SO_4. The results are recorded in tabular form thus:

ANALYSIS NO. I.

Substance blue, soluble in H_2O.

GROUP I. HCl	GROUP II. H_2S	GROUP III. $(NH_4)_2S$	GROUP IV. $(NH_4)_2CO_3$	GROUP V.
o	Brown precip., sol. in HNO_3. NH_4HO blue. K_4FeCy_6 brown. Cu.	o	o	o

ACIDS:

$BaCl_2$. White precipitate insol. in HCl . . . . H_2SO_4.

2. ANALYSIS OF $HgCl_2$.—This salt is likewise very soluble in H_2O. To one portion of the solution add HCl, and H_2S. The latter produces a black precipitate, insoluble in HNO_3, which indicates Hg, but a confirmatory test must be employed, which is to dissolve the precipitate in aqua regia and add $SnCl_2$. To a second portion

add some $BaCl_2$, which will cause no precipitate. To a third portion add $AgNO_3$, which produces a white precipitate of AgCl, insoluble in HNO_3, but soluble in NH_4HO, proving the presence of HCl.

3. ANALYSIS OF $FeSO_4$.—Acidify a portion of the solution with HCl, add a little H_2S to prove that no metals of the second group are present, and then $(NH_4)_2S$, which produces a black precipitate of FeS, which is treated as directed for Group III, page 277. To some of the original solution a drop of $K_6Fe_2Cy_{12}$ is added, when the blue color proves the presence of FeO.

4. ANALYSIS OF $Sr(NO_3)_2$.—Dissolve in water, test for Groups I, II, and III, which may occur as impurities, and then add $(NH_4)_2CO_3$. This white precipitate is filtered out and washed, then dissolved in acetic acid. To one portion add K_2CrO_4, when the absence of Ba is shown, and the Sr test may next be made, by adding NH_4HO and $CaSO_4$. The precipitate forms slowly. In the original solution no precipitate is formed by $BaCl_2$ or $AgNO_3$, and a careful test for HNO_3 is made with $FeSO_4$ and H_2SO_4, as described on page 279.

5. ANALYSIS OF $BaSO_4$.—This substance refuses to dissolve either in H_2O or in acids. It is boiled repeatedly with fresh quantities of Na_2CO_3 and filtered boiling hot, or fluxed with KNO_3 and Na_2CO_3. The filtrate contains Na_2SO_4 ; the residue is $BaCO_3$ and unaltered $BaSO_4$. The residue is dissolved in HCl and tested for Ba with K_2CrO_4 or $SrSO_4$ solution.

6. ANALYSIS OF A COIN.—A silver coin is dissolved in HNO_3, then diluted and the Ag precipitated with HCl as AgCl. From this metallic silver is precipitated on a piece of clean zinc placed in the precipitate and moistened with dilute H_2SO_4. In the blue filtrate will be found all the copper, which may be tested for as above. If instead of a silver coin, a nickel coin is used, HCl will give no precipitate, the Cu will be thrown down by H_2S, and the Ni by $(NH_4)_2S$. In analyzing compound substances, great care must be taken that *all* the metals of a certain group are precipitated before proceeding to the next, and for this purpose, after precipitating the Ag with HCl, a drop of HCl is added to the filtrate to ascertain whether any Ag remains in solution. Precipitates should also be well washed, but the wash-water is not added to the filtrate.

7. ANALYSIS OF TYPE METAL.—The type is cleaned and dissolved in nitric acid. If a white insoluble residue remains, it may contain both tin and antimony. To the filtered and diluted solution add just enough HCl to precipitate all the Pb, and filter *cold.* To the filtrate add H_2S as long as a precipitate is formed. Filter and wash thoroughly, digest with $(NH_4)_2S$ and filter. Dissolve the residue in HNO_3 and test for Pb and Bi. To the $(NH_4)_2S$ solution add excess of HCl to ascertain whether any portion of the precipitate was really dissolved. An orange red precipitate indicates Sb, which may be farther tested in a Marsh apparatus. The white residue, which was insoluble in HNO_3, is placed in a porcelain dish in contact with a slip of clean, smooth platinum foil, and heated with a little HCl. H_2O is then added, and a small fragment of Zn is put into the liquid, when Sn and Sb will be reduced to the metallic state. If the foil is blackened, Sb is present. The metallic residue is now dissolved in HCl, and $HgCl_2$ is added to ascertain if Sn is present.

8. Analysis of Tea-lead.—This substance, which is used by the Chinese for lining tea-chests, may be obtained of any retail grocer, or tea merchant. It consists of lead, tin, and copper, and the course of analysis differs little from that of type metal.

9. Analysis of Mixed Salts.—A mixture of $Pb(NO_3)_2$, $Bi(NO_3)_3$, $Co(NO_3)_2$, KNO_3 may be dissolved in water and the metals sought in the above order (viz. Pb, Bi, Co, K). In testing for acids the student will remember that if Pb was found among the metals, H_2SO_4 and HCl must have been absent, as either would have precipitated the lead. If the student forgets this and adds $BaCl_2$ to the solution, it will form a precipitate of $PbCl_2$, which he might mistake for $BaSO_4$, and hence incorrectly suppose H_2SO_4 to be present.

Mixtures of various other soluble salts should now be given out, such as $FeSO_4$, NaCl, $CuSO_4$, and NH_4Cl; gradually increasing the number of metals and acids to be sought.

10. Analysis of Lime-stone.—Dissolve any piece of marble or lime-stone in HCl. It will not be necessary to test for groups I and II; a small portion of the solution is tested with K_4FeCy_6 for iron. The alumina generally present is precipitated, along with the iron, by NH_4HO, after adding NH_4Cl, and is filtered out as rapidly as possible. In a small portion of the filtrate tests are made for Ba and Sr, which are of course absent, so that all the Ca may be precipitated by oxalate of ammonia. In the filtrate Mg will be found on adding Na_2HPO_4. The principal acid present is CO_2 as indicated by the effervescence with HCl when first dissolved. If a residue remained insoluble in HCl, it is probably SiO_2, or some silicate, and must be fluxed with Na_2CO_3 and K_2CO_3.

TABLE I.—SCHEME FOR

Add HCl *to Solution.*

PREC.

- Ag Pb Hg. Boil in H_2O
  - *Sol.* Pb. With K_2CrO_4, yellow.
  - *Prec.* Ag Hg. NH_4HO
    - *Sol.* Ag. With HNO_3, white.
    - *Prec.* Hg. Black.

FILTRATE.

- Add H_2S. PRECIPITATE.
  - Hg Pb Bi Cd Cu As Sb Sn Au Pt S. Digest with $(NH_4)_2S$
    - *Residue.* Hg Pb Bi Cd Cu. Boil in HNO_3.
      - *Res.* Hg. Test with Sn Cl_2.
      - *Sol.* Pb Bi Cd Cu. With H_2SO_4.
        - *Prec.* Pb. White.
        - *Sol.* Bi Cd Cu. NH_4HO
          - *Prec.* Bi. White.
          - *Sol.* Cd Cu. K Cy and H_2S
            - *Prec.* Cd. Yellow.
            - *Sol.* Cu. $HNO_3+K_4FeCy_6$ Red.
    - *Solution.* As Sb Sn Au Pt. In H apparatus with Zn & H_2SO_4
      - *Gas.* As Sb. Pass into Ag NO_3
        - *Sol.* As
        - *Res.* Sb and Ag
        - See page 276.
      - *Residue.* Sn Au Pt. Boil in HCl.
        - *Sol.* Sn. Test with Hg Cl_2.
        - *Res.* Au Pt. HCl and HNO_3. *Sol.*
          - Au. Sn Cl_2. Purple.
          - Pt. KCl. Yellow.

COMPLETE ANALYSIS.

FILTRATE.

Add H_2S.

FILTRATE

Add NH_4HO and $(NH_4)_2S$.

- *Prec.*
  - Co Ni Fe Cr Mn U Al Zn
  - Dilute HCl.
    - *Res.*
      - Co Ni.
      - Test with KNO_3 or borax bead, page 277.
    - *Sol.*
      - Fe Cr Mn U Al Zn
      - Boil in Na HO.
        - *Prec.*
          - Fe Cr Mn U
          - $(NH_4)_2CO_3$.
            - *Res.*
              - Fe Cr Mn
              - Fuse with KNO_3 and Na_2CO_3. Dissolve in H_2O.
                - *Res.*
                  - Fe
                  - KCyS. Red.
                - *Sol.*
                  - Cr
                    - $\bar{A}+Pb(NO_3)_2$. Yellow.
                  - Mn
                    - Green to red.
            - *Sol.*
              - U
              - $\bar{A}$ and K_4FeCy_6. Red.
        - *Sol.*
          - Al
            - Acidify with HCl and add NH_4HO, white.
          - Zn
            - H_2S. White.
- *Filtrate.*
  - Add $(NH_4)_2CO_3$.
    - *Prec.*
      - Ba Sr Ca
      - Dissolve in Acetic Acid
      - 2 Parts.
        - *I.*
          - Add K_2CrO_4
            - *Prec.*
              - Ba
              - Yellow precipitate.
            - *Sol.*
              - Sr
              - Add $CaSO_4$, white prec. in 30 min.
        - *II.*
          - Ca
          - H_2SO_4, filter; $NH_4HO+(NH_4)_2C_2O_4$ white.
    - *Filtrate.*
      - Mg K Na
      - 2 Parts.
        - *I.*
          - Mg
          - $NH_4HO+Na_2HPO_4$, white.
        - *II.*
          - Evaporate and test in flame; Na yellow, K violet.
      - Test for NH_3 in the original solution.

Name	Symbol	Atomic Weight	Molecular Weight	Specific Gravity	Specific Heat	Quantivalence	Electrical Character	Color	Fusing Point	Solvent
Aluminum	Al	27.5	$Al_2=55$	2.67	.2143	II or IV	+ 12	white	1292 F.	HCl
Antimony (Stibium)	Sb	122	$Sb_4=488$	6.71	.05077	III or V	— 16	bluish-white	812° F.	$HCl+HNO_3$
Arsenic	As	75	$As_4=300$	5.95	.08140	III or V	— 9	steel-gray	774° F.	HNO_3
Barium	Ba	137	$Ba_2=274$	1.85	......	II	+ 5	yellow	......	....
Bismuth	Bi	210	$Bi_4=840$	9.799	.03084	III or V	+ 28	reddish-white	507° F.	HNO_3
Boron.............	B	11	....	2.680	.2500	III	— 14	brown	....	K_2CO_3 hot
Bromine	Br	80	$Br_2=160$	3.187	.1060	I	— 5	brownish-red	45° F.	Ether
Cadmium	Cd	112	$Cd_2=224$	8.604	.05669	II	+ 24	bluish-white	4420° F.	H_2SO_4
Cæsium.............	Cs	133	$Cs_2=266$	unknown	......	I	+	....	....	....
Calcium	Ca	40	$Ca_2=80$	1.578	......	II	+ 7	light-yellow	....	HCl dil.
Carbon	C	12	$C_2=24$?	2.35	.02411	IV	— 15	black	infus.	insol.
Cerium.............	Ce	92	....	....	......	II	+ 15	....	....	....
Chlorine	Cl	35.5	$Cl_2=71$	2.435	.1214	I	— 4	green	....	H_2O
Chromium	Cr	52.5	$Cr_2=105$	6.810	......	II or IV	— 10	gray	....	HCl
Cobalt.............	Co	59	$Co_2=118$	8.95	.10696	II	+ 22	steel-gray	2000° F.	HCl
Copper (Cuprum)...	Cu	63.5	$Cu_2=127$	8.90	.09515	II	+ 30	red	1994° F.	H_2SO_4
Didymium...... ...	D	96	$D_2=192$	....	......	II	+ 16	....	....	....
Erbium	E	112	$E_2=224$	....	......	II	+ 11	....	....	....
Fluorine	F	19	$F_2=38$	1.31	......	I	— 3	....	....	....
Gallium............	Ga	......	....	4.7	(?)	....	+ (?)	white	85° F.	HCl
Glucinum	G	9.5	$G_2=19$	2.10	......	II	+ 9	....	....	....
Gold (Aurum)......	Au	197	$Au_2=394$	19.34	.03244	III	— 23	yellow	2015° F. (?)	$HCl+HNO_3$
Hydrogen	H	1	$H_2=2$	.069	3.4046	I	— 22	colorless	....	....
Indium......... ...	In	76	....	7.20	......	II	+ 26	white	349° F.	....
Iodine.............	I	127	$I_2=254$	4.94	.05412	I	— 6	bluish-black	225° F.	Ether
Iridium............	Ir	197	....	21.80	.03259	II or IV	— 25	white	O-H	insol.
Iron (Ferrum)......	Fe	56	... .	7.80	.11379	II or IV	+ 20	gray-white	1900 to 2900	Acids
Lanthanum..........	La	92	....	....	......	II	+ 17	....	....	....
Lead (Plumbum)....	Pb	207	....	11.36	.0314	II	+ 25	bluish-white	620° F.	HNO_3
Lithium	L	7	$L_2=14$	0.59	.9408	I	+ 4	white	324° F.	....
Magnesium	Mg	24	....	1.70	.2499	II	+ 8	white	446° F.	Acids
Manganese	Mn	55	....	8.00	.1217	II or IV	+ 18	gray-white	....	HCl
Mercury (Hydrargyrum)	Hg	200	....	13.596	.03332	II	— 30	white	—39° F.	HNO_3
Molybdenum.......	Mo	96	... .	8.64	......	VI	— 12	gray	....	HNO_3

Nickel	Ni	59	....	8.90	.10863	II	+ 21	white	2732°	HCl
Niobium	Nb	94	....	....	......	V	— 19	....	......	....
Nitrogen	N	14	N_2=28	.971	.2440	III or V	— 2	colorless	... .	....
Osmium...........	Os	199	....	21.400	.03063	II or IV	— 24	bluish-white	infus.	....
Oxygen	O	16	O_2=32	1.1056	.2182	II	—	colorless	....	....
Palladium.........	Pd	106	....	11—12	.05927	II or IV	— 29	white	4000° (?)	HNO_3
Phosphorus........	P	31	....	1.83	.1887	III or V	— 8	colorless	110°	CS_2
Platinum	Pt	197	....	21.53	.03243	II or IV	— 26	white	4591° F. (?)	HCl + HNO_3
Potassium (Kalium)	K	39	K_2=78	.865	.16956	I	+ 2	bluish-white	136°	....
Rhodium	Ro	104	....	11.00	.05803	II or IV	— 27	....	....	....
Rubidium..........	Rb	85	Rb_2=170	1.52	......	I	+ 1	white	101°	....
Ruthenium........	Ru	104	....	11—12	......	II or IV	— 28	....	....	....
Selenium...........	Se	79.5	Se_2=159	4.80	.0837	II or VI	— 7	brown	423°	CS_2
Silicon.............	Si	28	....	2.49	......	IV	— 21	brown	....	HNO_3 + HF
Silver (Argentum)..	Ag	108	Ag_2=216	10.50	.05701	I	+ 31	white	1873° F.	HNO_3
Sodium (Natrium)..	Na	23	Na_2=46	.972	.2934	I	+ 3	white	195°	....
Strontium..........	Sr	87.5	....	2.54	......	II	+ 6	pale-yellow	....	dil. HNO_3
Sulphur	S	32	S_2=64	2.00	.1776	II or VI	— 1	yellow	228°	CS_2
Tantalum...	Ta	182	....	10.78	......	V	— 18	....	....	....
Tellurium..........	Te	129	Te_2=258	6.65	.04737	II or VI	— 17	reddish-white	716°	H_2SO_4
Thallium	Tl	204	....	11.86	.03355	I or III	+ 23	white	554°	HNO_3
Thorium	Th	238	....	....	......	IV	+ 14	....	....	....
Tin (Stannum).....	Sn	118	....	7.3	.05623	II or IV	+ 27	white	442° F.	HCl
Titanium...........	Ti	50	....	5.3	......	II or IV	— 20	gray	....	HCl
Tungsten (Wolframium).......	W	184	....	17.50	.03342	VI	— 13	steel-gray	....	HNO_3
Uranium	U	120	....	18.40	......	III or V	+ 29	gray	....	....
Vanadium..........	V	51	....	3.64	......	III or V	— 11	....	....	....
Yttrium	Y	62	....	....	.....	II	+ 10	....	....	....
Zinc	Zn	65	....	7.15	.9555	II	+ 19	bluish-white	773° F.	H_2SO_4
Zirconium	Zr	89	....	4.15	......	IV	+ 13	black	.. ..	HF

NOTE.—The names of metals are printed in Roman, non-metals in *italics*.

The electrical character of an element is relative; although aluminum is positive towards a majority of the elements, it is negative towards 12 elements. Oxygen is the most electro-negative of all; sulphur is the most negative with one exception, hence marked —1. Cæsium is the most electro-positive of all and takes a plus sign. Silver is in middle numerically, there being 31 elements more strongly negative and 31 more strongly positive. Temperatures above 2000° F. are only approximate.

QUESTIONS FOR CLASS USE.

I.—INTRODUCTION.

Page 17.—Define chemistry. Illustrate. What is the distinction between organic and inorganic chemistry? Illustrate. What is an element? How many are known? Is it probable that all the elements have been discovered? Define chemical affinity. Illustrate. How does it act?

18.—How can we find what elements will combine? What is a compound? Are compounds like their elements? Illustrate. What is the action of heat? Of light? Of solution? Illustrate.

19.—State the principles upon which the elements are named. Illustrate each. What are chemical symbols? How are the symbols formed? Define atomic weight. Illustrate.

21.—State the four laws of the atomic theory. What is the molecular weight of a compound? What is a binary compound? Which element is placed first in reading the symbol of a compound? In writing? In reading the name? Name the three variations from these rules. What is the use of the terminations *-ide? -uret?* Define an oxide. Define the different compounds of O.

22.—What is an acid? Must an acid always be sour? Name one that is not (page 110). What is the test? How does the termination of an acid indicate its strength? What is the meaning of the prefix *per?—hypo?* How are the hydracids named?

23.—Define a base. An alkali. What is the reciprocal influence of the alkalies and the acids? Define a salt. How is it named? What is the difference between an *-ous* and an *-ic* salt? What is a formula?

24.—What signs are used? To what is the proportion of an element in any compound always equal? The weight of an element? State the proportion to be used in solving problems.

II.—INORGANIC CHEMISTRY.

1.—THE NON-METALS.

OXYGEN.—Give the symbol and atomic weight of oxygen. What is the meaning? What other element is an acid-former?

28.—Name the sources of O. Do these fractions indicate weight or volume? How is O prepared from potassium chlorate and manganese dioxide? Give the reaction. What becomes of the potassium chloride?

29.—What is the use of the black oxide of manganese? Define catalysis. Name the properties of O. What is oxidation? An oxide? Show that O is a supporter of combustion. What compounds are formed in these illustrations? What is an anhydride? An acid?

31–32.—Describe the destructive effects of the O in the air. What causes the decay of peaches? Why does not canned fruit decay? Describe the action of O on fuel. On the teeth. On impure water. On writing-ink. On red-hot iron. On damp knives and forks. How is river-water purified on a sea-voyage?

33.—By what means is the O carried through the system? What work does it perform in the body? Why is the blood in the arteries red and in the veins purple? What chemical processes are included by the chemist under the term oxidation?

34.—Does fire differ from decay? Is heat always produced by oxidation? Illustrate. Describe the body as a furnace.

35.—What is the chemical process of starvation? Why does unusual exercise cause one to breathe more rapidly? Why does running cause panting? Why do we need extra clothing when we sleep, even at midday, in the summer? How do hibernating animals illustrate this? How does a cold-blooded animal differ from a warm-blooded one? How does O give us strength?

36.—How are action and reaction equal in chemistry as in philosophy? What is potential force? Dynamic force? Show how O is constantly burning the body. Is there any part of the body that is permanent? Illustrate the rapidity of this change. Show the truth of the paradox—"*We live only as we die.*"

37.—Why do we need food and sleep? Show how O acts as a scavenger in nature. In what sense is O the sweeper of the body? Is this a useful provision?

38.—How much O does each adult need per day? Total amount used daily? What would be the result if the air were pure O? What objects would escape combustion? What is ozone? Where is it seen?

39.—Preparation? Properties? Test? Is it a valuable constituent of the air? What is antozone? Its characteristic?

NITROGEN.—Symbol and atomic weight? Why so called? Sources? Preparation?

42.—Properties? Why does a person drown in water? Would a person die in pure N? What is the peculiarity of the nitrogen compounds? What causes flesh to decompose so much more easily than wood? Does the N we take in at each breath do us any direct good or harm? Where do we get N to make our flesh?

43.—What use do plants make of the N they breathe in through their leaves? Describe the action of N and O in our stoves. Where do plants obtain N? State the main distinction between O and N. What is the office of the N in the air? Show that the proportion of O and N in the atmosphere gives us the golden mean.

44.—Symbol and molecular weight of nitric acid? Common name? Explain its occasional presence in the atmosphere. Preparation? Why is its symbol HNO_3 and not N_2O_5? Show that in its salts an acid takes the place of the H.

45.—Properties? Has it been obtained as a solid? How does it rank in strength? What color does it give to wood? Uses? Explain its oxidizing action. What is aqua regia? Describe the process of etching. The action of HNO_3 on Sn. What are the red fumes which pass off? How does HNO_3 illustrate the power of chemical affinity?

46.—Symbol and molecular weight of nitrous oxide? The common name? Preparation? Reaction? Properties? For what purpose has liquid nitrous oxide been used? (*Physics*, page 191.) What is the effect of nitrous oxide on the human system? State its use in surgical operations. Symbol and molecular weight of nitric oxide? Its preparation? Why is the gas in the jar colorless?

47.—What compound is formed? Properties of NO? What are the fumes which it forms in the air? Symbol and molecular weight of ammonia? Why so called? Its old name?

48.—What is aqua ammonia? Whence obtained? Give the reaction. Properties?

49.—Prove that H_3N is an alkali. What is its test? Its antidote? How liquefied? Define the nascent state.

HYDROGEN.—Symbol and atomic weight? Meaning of the name? Preparation?

51.—Reaction? What compound is formed? Properties? Is H a metal? *Ans.*—In all reactions it plays the part of a metal, and like most of the metals is electro-positive. The size of its atoms? Its levity? Will it destroy life? Effect on the voice? Use in filling balloons?

52.—What is the product of the combustion of H? What is the philosopher's lamp? What are the mixed gases? What is the cause of the report? Will the gases combine, if mixed? Describe the hydrogen gun.

54.—What compound is formed by the combustion of H? What becomes of the H_2O? What is the action of platinum sponge on a jet of H? What becomes of this force? Describe Dobereiner's lamp. Explain the heat produced by burning H.

55.—How are hydrogen tones produced? Explain.

WATER.—What is the freezing and the boiling point of water? How is the composition of water proved? Why does the blacksmith sprinkle water on his forge fire?

57–8.—What injury may a small quantity of H_2O do, if thrown on a fire? Explain. Can H_2O, then, be burned? Show that electrical force is latent in water. What becomes of this force? Illustrate the abundance of H_2O in the animal world. Vegetable world. Mineral world.

59.—Why will blue vitriol lose its color if heated? What is "burnt alum?" Water of crystallization? Show the adaptation of H_2O as a solvent. What water is the purest? Why does rain-water taste so insipid?

60.—Is river water a healthy drink? What is hard water? Soft water? Why does the hardness of water vary in different places? Is hard water healthful? How may we detect organic matter in H_2O? What minerals are most common in water? What is the "fur" in a tea-kettle? Why does soap curdle in hard water?

61–2.—What is the cause of the tonic influence of the sea-breeze? How could Salt Lake be freshened? What is the use of the air in H_2O? How do fish breathe? Why does the air in water contain so much O? Why is boiled water so insipid? Give some of the paradoxes of water. Name the various uses of water. (*Physics*, p. 201.)

CARBON.—Symbol and atomic weight? Illustrate the abundance

of C. Is it more characteristic of the vegetable than of the mineral kingdom? What are its three forms?

65.—Proof of these allotropic states? What is an allotropic condition? What is the diamond? Properties? Has it ever been made artificially? What is a carat?

66.—How is the diamond ground? Describe the three modes of cutting. What gives the diamond its value?

67.—Common name for graphite? Origin? Uses? Describe the process of making a lead-pencil. What is a black-lead crucible? What is British Lustre? Gas carbon? How is charcoal made?

68.—What is the chemical change? Illustrate the durability of charcoal.

69.—Its property of absorbing gases. Its preservative effects. Its filtering properties. (*Physics*, page 17.) What do you mean by the deoxidizing or reducing action of C? Application to the arts?

70.—What is soot? What causes the burning of chimneys? Does this occur oftener when wood than when coal is used as fuel? How is lampblack made? Uses? Fitness for printing? What can you say about ancient MSS.? How is boneblack made? Uses? How is sugar refined?

71.—Describe the formation of coal. Difference between bituminous and anthracite coal. Why is coal found in layers, with slate, etc., between? Why is the coal hidden in the earth? What proof have we that coal is of vegetable origin?

72.—What is coke? Uses? Describe the formation of peat. Uses? What is muck? Use? Name some of the diverse properties and uses of C.

CARBONIC ANHYDRIDE.—Symbol and molecular weight? Sources? How it is constantly formed?

74.—Preparation? Reaction?

75.—Test? What causes the pellicle on lime-water? What does this show? Prove that we exhale CO_2. Give the properties of CO_2.

76.—Prove that CO_2 is heavier than air. A non-supporter of combustion. That it contains C. What test should be employed before descending into a deep well or an old cellar? How can you remove the foul air?

77.—Tell about the Grotto del Cane. Is CO_2 directly poisonous? What is choke-damp? Fire-damp? Which is more dreaded?

78.—Has CO_2 been used in extinguishing fires? Tell about the absorption of CO_2 by H_2O. What is soda-water?

79.—How is CO_2 liquefied? Why does the liquid acid solidify when exposed to the air? What principle in philosophy does this illustrate? How low a degree of cold has been produced in this manner? Describe the need of ventilation. How is the air expired from our lungs made useful?

80.—Is a single opening sufficient to ventilate a room? What practical application do you make of this subject? Symbol and molecular weight of carbonic oxide? Properties?

81.—Where do we often see it? How is CO formed in a coal-fire? Practical importance of this fact? What causes the unpleasant odor of coal-gas? Symbol and molecular weight of light carburetted hydrogen? Properties? How is it formed?

82.—Name the places where it is found in great quantities. Symbol and molecular weight of heavy carburetted hydrogen? Properties?

83.—What gases mainly compose coal-gas? Which is the most valuable? Describe the manufacture. Is the odor beneficial? Is coal-gas explosive? Why is the jet flat? When we turn the gas very low, or the supply is insufficient, why is the flame blue? Symbol and molecular weight of cyanogen? Meaning of the name?

84.—Preparation? What are its compounds called? What is the yellow prussiate of potash? The red? *Ans.*—The ferricyanide, K_3FeCy_6. Properties of Cy? What is a compound radical? Symbol of hydrocyanic acid? Common name? Where found? Antidote? What are the fulminates? How are gun-caps made?

COMBUSTION.—Define. What is a combustible? A supporter of combustion? (The difference between these two is nicely shown in the experiment with H on p. 52.) A burnt body? *Ans.*—A body which has combined with O.—*Example:* a stone, water. Upon what does the amount of heat produced by combustion depend? The intensity? Why do we need a draught to a stove? What is meant by the igniting point of a substance? Does combustion, in its chemical sense, commence before the fuel catches fire? Why do we use "kindlings" in starting a fire? Why can we light pitch-pine so easily? What are hydrocarbons? What are the ordinary products of combustion? What causes the dripping of stove-pipes? What are the ashes? Why does fresh fuel produce a flame?

86.—Show how wisely C is adapted for a fuel. What would be the effect if CO_2 were not a gas? Define flame. Describe the burning of a candle.

87.—Show that flame is hollow. What causes the light? Why is the flame blue at the bottom? Products of combustion? Tests? Why does the wick turn black?

88.—What causes the coal at the end of the wick? Why does snuffing brighten the light? Why does a draft of air, or a sudden movement of the candle, cause a deposit of soot? Why is the flame of a candle or lamp red, or yellow? *Ans.*—Because the heat is not sufficient to cause the carbon to emit all the rays of the spectrum. Use of plaited wicks? Object of a chimney to a lamp?

89.—A flat wick? Advantage of an Argand lamp? What is the film which gathers on the chimney when the lamp is first lighted? Why does this soon disappear? Why do tar, spirits of turpentine, etc., burn with much smoke? Why does alcohol give much heat and no smoke? Describe Davy's safety-lamp. Illustrate this by a wire gauze over the flame of a candle.

90.—Describe Bunsen's burner. Why does it give great heat, little light, and no smoke? Describe the oxy-hydrogen blow-pipe.

91.—Why does it give great heat and little light? What is the calcium light?

92.—Describe the mouth blow-pipe. The three parts in the blow-pipe flame. What is the reducing flame?

93.—The oxidizing flame? Why does blowing on a candle-flame extinguish it? Why does water put out a fire? Give illustrations of spontaneous combustion.

THE ATMOSPHERE.—Name the constituents. Proportion. State the comparison. What is the law of diffusion?

97.—What effect does this have on the air? Is the air a chemical compound? Illustrate. Has each constituent a special use? Name the uses of O. Of CO_2.

98.—Explain the chemical change which takes place in the leaf. What force separates the C from the O? What is the influence of house-plants upon the atmosphere of a room? What do you say of the exact balance kept between the wants of animals and plants?

100.—What relation exists between animals and plants? Which gathers and which spends the solar force? Which performs the office of reduction? Which that of oxidation? How is the solar force set free? What is the use of the watery vapor in the air?

101.—Which of the constituents are permanent? Is this a wise provision? Why ought the vapor to be easily changed to the liquid form? What effect does this permanence have upon sound?

THE HALOGENS.—Name them. Symbols and atomic weights. Compare the halogens with each other. What compounds do they form? Why is chlorine so called? Source?

103.—Preparation? Reaction? Properties? What action does Cl have on phosphorus, arsenic, etc.? Why does a solution of the gas soon become acid? What is its action on organic bodies? On turpentine? On printers' ink? Describe the chemical change in domestic bleaching.

104.—The method of bleaching on a large scale. What is the advantage of using Cl over other disinfectants? How may the gas be set free?

105.—How are hospitals purified? What mixture would liberate Cl in the greatest quantities? Symbol and molecular weight of hydiochloric acid? Common name? Preparation? Reaction? Properties? What is muriatic acid? What are its compounds termed? Tests? What is nitro-muriatic acid?

106.—Symbol and molecular weight of chloride of lime? Uses? Symbol and molecular weight of calcium chloride? Preparation? Peculiar property? Tell what you can about bromine. Its uses. What is the peculiarity of fluorine? Source? What acid does it form? For what is this acid noted? Describe the process of etching with HF.

107.—Why is not HF kept in ordinary bottles? Is it dangerous to use? Why is iodine so called? Preparation? Properties? For what are its compounds noted? How may its stains be removed? Test? Use in medicine?

BORON.—Symbol and atomic weight? Source? Describe the scene in Tuscany where it is found. Process of manufacture.

109.—Symbol and molecular weight of borax? Uses in soldering, and in softening hard water?

SILICON.—Symbol and atomic weight? Source? Common names? What gems does it form? What is sand? Properties?

110.—Why is it called an acid? Is silica soluble in H_2O? How does it get into plants? In what plants is it found? Explain the process of petrifaction. What is said of the antiquity of glass? Pliny's story of its origin? What is said of its value in the twelfth century?

111.—Name the four varieties of glass and the composition of each. What are the essential ingredients of glass? How is glass colored? Name the oxides used. Why is flint-glass so called? How is glass annealed? Describe the Prince Rupert's drop.

112.—How are Venetian balls made? Tubes? Beads?

SULPHUR.—Symbol and atomic weight? Sources? What is the principle of hair-dyes? Why do eggs tarnish silver spoons? What is the difference between brimstone and flowers of sulphur? Properties? Solvent? **Four** allotropic forms? Describe the amorphous state.

114.—Uses of S? Symbol and molecular weight of sulphurous anhydride? Where is it familiar? What are its compounds called? Uses in bleaching? Why are new flannels liable to turn yellow when washed? Symbol and molecular weight of sulphuric anhydride? By what other name is it known? Preparation? Properties? Why is Nordhausen acid so called?

115–16.—Symbol and molecular weight of sulphuric acid? Common name? State its importance. What are its compounds called? Illustrate the making of H_2SO_4. Describe its manufacture. Reaction.

117.—Properties? What especial property? Illustrate. Its strength? Color of its stain on cloth? How removed? On wood? Cause of this action? Test? Symbol and molecular weight of hydrogen sulphide? Where is it found?

118.—Preparation? Reaction? Properties? Use? Color of these precipitates? Test? Symbol and molecular weight of carbon sulphide? Preparation? Properties? Uses? How does it illustrate the force of chemical affinity?

PHOSPHORUS.—Symbol and atomic weight? Why so called?

120.—Source? In what parts of the body, and in what forms, is it found? Preparation? Properties? Caution to be observed? Is phosphorus poisonous? What is the product of its combustion?

121.—Describe the amorphous form of phosphorus. What is the principal use of phosphorus? Describe the making of the lucifer match. The safety match. What compounds are formed in the burning of a match?

122.—What is phosphorescence? Its cause? Symbol and molecular weight of hydrogen phosphide? Source? Preparation? Properties?

ARSENIC.—Symbol and atomic weight? Common name? Test? What is commonly sold as arsenic or ratsbane? Preparation of arsenious acid? Properties? What can you tell of its antiseptic property? Antidotes? Describe Marsh's test. How can the As be distinguished from Sb?* What is said of arsenic eating?

2.—THE METALS.

POTASSIUM.—Symbol and atomic weight? History of its discovery? Source? How do we get our supply? Preparation?

127.—Properties? How must it be kept? Reaction when thrown on H_2O? Symbol and molecular weight of potash? Of potassium hydrate? Properties? Its feel? Its affinity for H_2O and CO_2? Uses? Symbol and molecular weight of potassium carbonate? Common name?

128.—Preparation? What part of the tree furnishes the most potash? What is the derivation of the word? Symbol and molecular weight of hydrogen potassium carbonate? Common names? Preparation? Define a rational formula. Illustrate. An empirical formula? Illustrate. What is a neutral salt? An acid salt? A dibasic acid?

129.—Symbol and molecular weight of potassium nitrate? Common names? Where is it found? How is it prepared artificially? How much water would be required to dissolve a pound of this salt? Properties? Uses? What is the composition of gunpowder? Cause of its explosive force?

130.—Uses of potassium chlorate? Potassium bichromate? Composition of fire-works?

SODIUM.—Symbol and atomic weight? Source? What proportion does it form of common salt? What element does it resemble?

131.—Reaction when thrown on water? What compound is formed? Test? Symbol and molecular weight of common salt? What use does it subserve in the body? Is salt abundant? Describe the manufacture.

132.—What is solar salt? Describe the "hopper-shape" crystal. Is it best to heat the water for dissolving salt? What is a saturated solution?

133.—Symbol and molecular weight of sodium sulphate? Common name? Preparation? Reaction? What curious property has this salt? Why will the dropping in of a crystal cause solidification?

* Antimony is more likely to be mistaken for arsenic than for any other metal. The crust which is formed by decomposing antimoniuretted hydrogen in Marsh's apparatus does not yield octahedra, when sublimed in a tube with air, but prisms. The metal is also easily soluble in yellow ammonium sulphide, which is nearly without effect upon arsenical crusts.—MILLER.

Symbol and molecular weight of sodium carbonate? Common names? Why called carbonate of soda?

134.—Describe its manufacture. Why will Na_2SO_4 soften hard water? Symbol and molecular weight of hydrogen sodium carbonate? Common name? Why called bicarbonate of soda? Preparation? Use? What is an empirical formula? A rational formula? Give the theory of ammonium. How is the symbol H_4N,HO obtained? What is a compound radical? A compound halogen?

135.—Symbol and molecular weight of ammonium chloride? Preparation? Uses? Symbol and molecular weight of ammonium carbonate? Common names? Uses? Symbol and molecular weight of ammonium nitrate? Preparation? Uses? What is the sodium amalgam?

CALCIUM.—Symbol and atomic weight? Source? In what part of the body is it found? In what form do we commonly see it? Symbol and molecular weight of lime? Preparation? Describe a lime-kiln. Properties of CaO? Test?

137.—What is the difference between "water-slacked" and "air-slacked" lime? Uses? What is whitewash? Concrete? Hard-finish? Calcimining? Theory of the hardening of mortar? Why are newly-plastered walls so damp? Will mortar harden if protected from the air? Action of lime on the soil? Will it not lose its beneficial effect after a time? Should it be applied to a compost heap? How can this waste be avoided? How would you test for the escaping H_3N? Action of lime on copperas? How does the copperas get in the soil?

138.—Uses of lime? Symbol and molecular weight of carbonate of lime? Source? How are stalactites and stalagmites formed?

139.—What is petrified moss? Whiting? Marble? Chalk? Marl? Symbol and molecular weight of calcium sulphate? Common names? What is plaster of Paris? Why does plaster of Paris harden, if moistened? *Ans.*—Because it absorbs water again. Uses? What is plaster? How prepared for use as a fertilizer? *Ans.*—It is ground into a fine powder. Tell the story of Franklin.

140.—What is the difference between sulphate and sulphite of lime? Symbol and molecular weight of phosphate of lime? What is the superphosphate? Use? Uses of the salts of barium and strontium? What is heavy spar? Barytes?

MAGNESIUM.—Symbol and atomic weight? Source? How can you tell if a stone contains Mg? *Ans.*—It generally has a soapy feel. Properties? For what is it noted? Name the two classes of rays contained especially in its light (*Physics*, page 163). For what purposes do the colorific rays adapt it? Does it contain heat-rays also? *Ans.*—Very few. Product of its combustion? Symbol and molecular weight of magnesium carbonate? Magnesium sulphate? Common name?

ALUMINUM.—Symbol and atomic weight? Common name? Source? Properties? Solvent? What can you say of its abundance and probable usefulness? What is alumina? What crystals and gems does it form? What is emery? Symbol and molecular weight of silicate of alumina? Common name? Source?

144.—Use in the soil? In the arts? What is ochre? Fuller's earth? Explain the process of glazing pottery ware. What is the salt glaze? The litharge glaze? What objection to the latter? What gives color to brick? What is the peculiarity of white brick? How is alum made?

145.—Name the different kinds of alum. Which kind is the common commercial alum? Use of alum in dyeing? How are alum crystals made? *Ans.*—They are obtained by suspending threads in a saturated solution of this salt. In this manner alum baskets, bouquets, etc., are formed of any desired color. What is spectrum analysis? Is it a reliable test? Illustrate its delicacy? What is the spectroscope?

IRON.—Symbol and atomic weight? Tell what you can of its value to the world. How is its use a symbol of a nation's progress?

149.—State how its value is enhanced by labor. Name the sources of iron. Common ores. Describe the process of smelting iron ore. Why is hot air used for the blast? Reaction of the lime?

151.—What becomes of the O in the ore? Origin of the term "pig-iron?" Name the varieties of iron. Difference between them. What is cast-iron? Its properties? Uses? What exception adapts iron for use in castings? What does this teach us? What is chilled iron? Wrought iron?

152.—Preparation? Effect of jarring? How is Fe tempered? Illustrate its malleability. What is steel? Preparation? In making steel tools, how does the workman judge of the temper? How are cheap knives made?

153.—Describe Bessemer's process. Cause of the changing colors often seen in the scum over standing water?

154.—Name the different oxides of iron. Give the symbol of each. Where is each found? Origin of colored sand? What peculiar property is possessed by the ferric oxide and ferric hydrate? What is iron carbonate? By what name is it known?

155. — Cause of the ferruginous deposit around chalybeate springs? Symbol and molecular weight of iron bisulphide? Common names? Of ferrous sulphate? Preparation? Uses? What is chameleon mineral?

ZINC.—Symbol and atomic weight? Source? Preparation? Reaction? Is it malleable? Will it oxidize in the air? Uses? What is philosopher's wool? What is galvanized iron? Are water-pipes made of this material safe? Symbol and molecular weight of zinc oxide? Use? Symbol and molecular weight of zinc sulphate? Use?

TIN.—Symbol and atomic weight? Where found? Properties? What is the "tin cry?" What is common tin-ware? Action of HNO_3 on Sn? What can you say of the manufacture of pins?

COPPER.—Symbol and atomic weight? Where found? Antiquity of the mines? What is malachite? Properties of Cu? Color of its vapor? How tempered? Test? What is verdigris? Carbonate of copper? Black oxide of copper? What is the danger of using a copper kettle? Solvent of Cu? Test? Symbol and molecular weight of copper sulphate? Common name? Uses?

LEAD.—Symbol and atomic weight? Source? Preparation? Properties? Its effect on the human system? On water? Is there more danger with hard, or with soft water? What precaution should always be used with lead pipes? What is the test of lead? What is "litharge?" Its uses? "Red-lead?" Its uses? What is "white-lead?" Describe its manufacture. With what is it adulterated? What is "sugar of lead?" Properties? Antidote? Explain the formation of the lead-tree.

GOLD.—Symbol and atomic weight? Source? Preparation? What is an amalgam? Quartation? Properties? Solvent? Process of making gold-leaf?

SILVER.—Symbol and atomic weight? Source? Preparation, 1, from the sulphide; 2, horn-silver; 3, lead? Describe the process of reduction at the West. What is cupellation? Properties? Solvent? Test? What is the common name of nitrate of silver? What is its action on the flesh? How may its stain be removed? Uses? Of what are hair-dyes and indelible inks made? Describe the process of Daguerreotyping. Photography.

PLATINUM.—Symbol and atomic weight? Source? Preparation?

Properties? Uses? Why is iridium so named? What is iridosmine? How is platinum wire made?

MERCURY.—Symbol and atomic weight? Common name? Why so called? Source? Preparation? Properties? Uses? Action on the human system? Process of silvering mirrors? What is blue-pill? Mercurial ointment? Symbol of mercuric oxide? Mercurous chloride? Mercuric chloride? Common name? Uses? Properties?

THE ALLOYS.—What is an alloy? What peculiarity with regard to the melting point? Of what is type-metal made? Pewter? Britannia? Brass? German silver? Solder? Fusible metal? Bronze? How is gold soldered? Silver? Copper? What is the principle? What are the constituents of gold coin? Silver coin? What is the meaning of the term carat? How are shot manufactured? How are they sorted? What is oreide? Aluminum bronze? Compare the properties of the metals with regard to, 1, oxidation; 2, density; 3, melting point; 4, color; 5, malleability; 6, brittleness; 7, tenacity; 8, special properties.

III.—ORGANIC CHEMISTRY.

INTRODUCTION.—Why must matter be organized? What is the office of plants?

182.—What is the difference between organic and inorganic bodies? Illustrate each of the four distinctions.

183.—What is the number of carbon compounds? Define isomerism. Illustrate. What is the cause? Allotropism? Illustrate.

STARCH.—Symbol and molecular weight? Sources? Use in the plant? Why stored in that form? Appearance under the microscope? Preparation? Properties? What is dextrine? Test of starch? Varieties? What is gum? Composition? Mucilage? Is it soluble in water? What is pectose? Pectin?

WOODY FIBRE.—Symbol and molecular weight? What is the composition of wood? Name the various forms of cellulin. Illustrate the wonders of secretion. State the uses of woody fibre. The making of paper. Paper-parchment. Linen. Cotton. Gun-cotton [$C_6H_7(NO_2)_3O_5$]. Collodion. Its uses. Cane-sugar. How is sugar refined? Difference between loaf and granulated sugar? Describe a centrifugal machine. What is terra alba? Use? Of what are gum-drops made? Rock-candy? What is caramel? Use? Symbol and molecular weight of grape-sugar? Source?

Sweetening power? How is sugar made from starch? How does the oil of vitriol act? How do jellies, preserves, etc., "candy?" Why are dextrose and levulose so named?

FERMENTATION.—Cause? Does it ever take place spontaneously? How does the yeast act? What change takes place in the alcoholic fermentation? The acetic? Describe the formation of yeast. The making of malt. Yeast cakes. The varieties of fermentation. What is gluten? How does it act? What is diastase? Describe the brewing of beer. Why is lager beer so called? Describe the making of wine. What is the difference between a dry, a sweet, and an effervescing wine? Cause of the flavor? State the proportion of alcohol in common liquors. How is brandy made? Rum? Whisky? Gin? Describe the apparatus used for distillation. Symbol and molecular weight of alcohol? What is said of its affinity for water? What is absolute alcohol? Name the uses of alcohol in the arts. Effects of alcohol on the human system. Symbol and molecular weight of ether? Is it properly called sulphuric ether? (See p. 203.) Preparation? Properties? Uses? Preparation of chloroform? Properties? Uses? What is chloral? Chloral hydrate? Properties? Uses? Symbol and molecular weight of acetic acid? What is the glacial acid? Preparation of vinegar? What causes the working of cider? What change takes place? Properties of acetic acid? Use? What causes the "working" of preserves? What is aldehyde?

ORGANIC RADICALS.—What is a radical? A homologous series? Name the terms of the marsh-gas series. What is ethyl, methyl, etc.? By what other name is marsh-gas known? Describe the formation of the alcohols. By what other name is common alcohol known? State the formation of the aldehydes and acids. The ethers? The compound ammonias. The salts of the radicals. Uses.

DESTRUCTIVE DISTILLATION.—What change takes place in the decay of wood? Effect upon the soil? What change takes place in the distillation of wood? Why is it called "destructive?" What is pyroligneous acid? Use? Creosote? Properties? Uses? Paraffine? Properties? Uses? How is tar made? What are the products of the distillation of coal-tar? Properties of carbolic acid? What are the picrates? Uses? What is benzole? Benzine? Properties? Uses? What is phenyl alcohol? What is nitro-benzole? What are the coal-dyes? Give an account of their discovery and properties. What is naphtha? Naphthaline? Anthracene?

Alizarin? Dead-oil? Uses? Petroleum? How formed? Describe its distillation. The rectification of kerosene. Danger of kerosene explosions. The test given by Dr. Nichols. How is bitumen formed? Describe Tar Lake. What is "Greek fire?"

THE ORGANIC ACIDS.—Where is oxalic acid found? Preparation? Properties? Antidote? Uses? Where is tartaric acid found? Preparation? What is cream of tartar? Tartar emetic? Rochelle salt? Seidlitz powders? Where is malic acid found? Citric? Tannic? Name its varieties. What are nut-galls? Properties of tannin? Describe the process of tanning. How is leather blackened? How is ink made? Why does writing-fluid darken by exposure to the air? What is gallic acid? Pyrogallic? Use?

THE ORGANIC BASES.—Sources? What is opium? Preparation? Uses? Laudanum? Paregoric? Danger of opium-eating? What is morphine? Use? Quinine? Use? Nicotine? Properties? Strychnine? Properties? The chromatic test? Name the active principle of tea and coffee. What substances are found in tea? In coffee? Describe the process of tea-raising. Of making black tea. Green tea.

ORGANIC COLORING PRINCIPLES.—Source? What is an adjective color? A substantive color? A mordant? The process of dyeing? Of calico printing? What is madder? Its coloring principle? Cochineal? Use? Brazil-wood? Use? Indigo? Preparation? White indigo? Logwood? Litmus? Leaf-green?

OILS AND FATS.—Name the two classes. What is the difference between them? What is the composition of the fatty bodies? Illustrate. What is glycerin? Uses? Nitro-glycerin? Illustrate the formation of soap. What is the reaction? Difference between hard and soft soap? What is the cause of the curdling of soap in hard water? Describe the cleansing action of soap. What is saponification? How is stearin made? What are adamantine candles? Is wax of animal or vegetable origin? How is it bleached? What is a drying oil? Boiled oil? Putty? Printers' ink? Cod-liver oil? Croton oil? Castor oil? Sweet oil? Uses? Sources of the volatile oils? Preparation? Composition? Name the three classes. Illustrate each. What is oil of turpentine? Rosin? Camphene? Camphor? Preparation? Properties?

RESINS AND BALSAMS.—What is the difference between a resin and a balsam? Illustrate. Source? Properties? Uses? What is rosin? Preparation? Uses? Lac? Source? Preparation? Shellac? Sealing-wax? Gum Benzoin? Uses? Amber? Origin?

Properties? Uses? India-rubber? Source? Properties? Uses? What is vulcanized rubber? Properties? Gutta-percha? Uses?

ALBUMINOUS BODIES.*—Name them. What is their composition? What is albumen? Source? Properties? Casein? Why does milk curdle? Action of rennet? Why does cream rise on milk? Describe the souring of milk. What is gelatin? Glue? Isinglass? Size? Fibrin? Properties? Gluten? Legumin? Putrefaction? Cause? Why does salt preserve meat?

DOMESTIC CHEMISTRY.—Describe the chemical changes which take place in making bread. What is stale bread? Why is it dry? How is aerated bread made? Why is bread ever sour? How are griddle-cakes raised? Biscuit? What are baking-powders? Action of soda and HCl? Of sal-volatile? How is bread changed by toasting? How are potatoes changed by cooking?

* Notice here the wise provision of nature. Nitrogen, slow and sluggish when uncombined, is fitted to dilute the air; while N, restless and uneasy when combined, is equally adapted to form unstable compounds of food, to carry force into our bodies and there to quickly set it free. Oxygen, when free, is active, eager, and ready to search the nooks and crannies of the capillaries; but when once it combines with a substance, takes it for better or for worse, and forms the stablest of compounds. We find nitrogen compounds in the animal and vegetable worlds, ready for use where they are needed, in our muscles. Oxygen compounds are abundant in the mineral world, and stored in the seeds of plants, at hand to give form to the more permanent parts of the body. Such profound relations, such nice adaptations of our bodies to the world around, give us glimpses of a creative skill worthy our noblest thought and highest admiration.

CHEMICAL APPARATUS.

This Set is adequate for the performance of the *leading* experiments in any text-book. *We would call special attention to the extremely low price at which it is offered.*

PRICE, $15.

Alcohol.
Acid, Sulphuric.
" Nitric.
" Hydrochloric.
" Arsenious.
" Oxalic.
" Tartaric.
Ammonium Chloride.
" Nitrate.
" Sulphide.
Ammonia Water.
Antimony (Metallic).
Barium Chloride.
" Nitrate.
Bone Black.
Calcium Fluoride.
" Sulphate.
Copper Sulphate.
Carbon Disulphide.
Ether.
Ferrous Sulphide.
" Sulphate.
Gun Cotton.
Iodine.
Lead Acetate.
Litmus.
Mercury.
Magnesium Ribbon.
" Monoxide.
Manganese Dioxide.
Nut Galls.
Phosphorus.
Potassium.
" Ferrocyanide.
" Chlorate.
" Hydrate.
" Nitrate.
" Bichromate.
Strontium Chloride.
" Nitrate.
Sulphur.
Silver Nitrate.
Sodium (Metallic.)
" Biborate.
" Carbonate.
" Sulphate.
Blowpipe.
Deflagrating Spoon.
Evaporating Dish.
Evolution Flask.
Filters.
File.
Funnel.
Graduate.
Glass Tubing.
Lead Dish.
Nest of Crucibles.
Retort.
Rubber Tubing.
Spirit Lamp.
Test Tubes.
Tripod.
Wedgewood Mortar.

We would recommend the above to such schools as have not the means to purchase the more comprehensive set.

INDEX.

This Index includes the Notes as well as the Text.

www.ingramcontent.com/pod-product-compliance
Lightning Source LLC
LaVergne TN
LVHW020238110826
845151LV00003B/956